Food and Identity in a Globalising World

Series Editors
Atsuko Ichijo
Department of Politics and IR
Kingston University
Kingston-Upon-Thames, UK

Ronald Ranta
Department of Politics and IR
Kingston University, UK

This series aims to overcome the current fragmented nature of the study of food by encouraging interdisciplinary studies of food and serving as a meeting place for a diverse range of scholars and practitioners who are interested in various aspects of food. By encouraging new original, innovative and critical thinking in the field and engaging with the main debates and controversies, and by bringing together the various disciplines that constitute food studies, such as, sociology, anthropology, politics and geography, the series will serve as a valuable source for researchers, practitioners, and students. There will a focus on identities and food; issues such as gastrodiplomacy, settler colonialism, gender, migration and diaspora, and food and social media, while at the same time promoting an inter- and trans-disciplinary approach.

More information about this series at
http://www.palgrave.com/gp/series/16371

Brandi Simpson Miller

Food and Identity in Nineteenth and Twentieth Century Ghana

Food, Fights, and Regionalism

palgrave
macmillan

Brandi Simpson Miller
Lane Center for Social and Racial Equity
Wesleyan College
Macon, GA, USA

ISSN 2662-270X ISSN 2662-2718 (electronic)
Food and Identity in a Globalising World
ISBN 978-3-030-88402-4 ISBN 978-3-030-88403-1 (eBook)
https://doi.org/10.1007/978-3-030-88403-1

Cover illustration: LindaHughesPhotography / Alamy Stock Photo

This Palgrave Macmillan imprint is published by the registered company Springer Nature Switzerland AG.
The registered company address is: Gewerbestrasse 11, 6330 Cham, Switzerland

Preface

As a woman of colour in the field of history, I have been grappling with questions about the colonizing effects of western perspectives on the teaching of history. Teaching history often means directly encountering narrative violence and the hegemony of whiteness. Instead of relating the ideas that influence my inclusive contribution to secondary and tertiary teaching and scholarship, I wish to reference episodes that authentically articulate my engagement with reparative practices in the teaching and research of history.

When I was four years old, my family moved to Duisburg in what was then West Germany. During this time, my mother and father did their best to embrace the local culture, adopting and serving German food in the tiny apartment we called home. Kartoffelpuffern, brötchen, and marzipan were some of my favourite dishes. When I was ten, we returned to the United States, and I became aware of a new social identity I had inadvertently assumed. I was now Black or African American. I had no idea what this meant, having never been addressed this way in Germany. This social death and rebirth into a fraught new identity has traumatized me in ways I still struggle with today. I still identify with the alienation of being a foreign student having witnessed my older siblings' adjustment to life first in Germany and then the reverse culture shock of adjusting to life in the midwestern state of Minnesota.

My experience as both a student and as a history teacher at both the primary and secondary levels has demonstrated that although African history may be part of the standards, teachers do not instruct students in this history as graduation tests do not hold them accountable for this information. My own experience living in the United States was one characterized by a profound sense of loss, bewilderment, and a persistent sensation of being unmoored which was exacerbated by the absence of any information about people who looked like me.

My research is an attempt to learn about and reconcile my own identity as half American and half Nigerian, and to learn about how people in Ghana conceptualized theirs. My work began with a curiosity about how new world foods affected the social history of West Africa. I learned a bit of what it means to be an African in Africa. This reality differs considerably from the bits of information about Africa I was able to absorb from popular culture in the west. My mission is to promote teaching and learning about the old world of Africa at the primary and secondary levels, to dispel the profound ignorance that most westerners have about the cultural and social history there. My commitment to social change in the United States is reflected in the current direction of my research, which addresses the erasure of the black experience in the national imagination. African history relates to our story as Americans and is a history which we all carry and sustain in our art, music, and foodways. My innovative food studies approach challenges black erasure by providing an historical context that engages scholars in several different fields, while at the same time appealing to the layperson's interest in food and cooking, which has grown steadily over the last three decades. It is my sincere hope that I was successful in introducing Ghanaian perspectives into this work on West African foodways and that it will be a meaningful contribution to the fields of food studies, African history, and comparative studies.

I chose to write about Ghanaian food history quite by accident when I began my Masters course in World History. The very last semester of the course I took a class on West African History as a fulfilment of the requirements. In a class at Georgia State about Ghana, I learned about the cash crops farmed there and began to wonder about what people ate there. There was very little written about this, so I decided that what people ate in Ghana and its relationship to ideas about national identity would by my Masters thesis topic. This was at a time where the work of Alfred Crosby was experiencing a revival and the work of Sydney Mintz began to expand historical inquiry about food and its relationship to the environment and to the political organization of humans on this planet. Further, scholars like Richard Wilk and Jeffrey Pilcher were popularizing research on the connection between food and nationalism in world history.

I discovered a lack of serious intellectual engagement with, and a weakness and inconsistencies in the scholarly knowledge of the culinary traditions of West Africa. I became intrigued by the challenge of researching and writing about this part of the world, part of the appeal of which was the interdisciplinary approach the work demanded.

With support from both the West African Research Centre at Boston University and the Royal Historical Society in London, I researched the topic of this book by travelling to Ghana where I toured ports, visited government ministers, interviewed some of the owners of cooking schools, spoke with chefs and cooks at every type of eatery, and observed the processing of food all around me. I also undertook archival research, reviewing newspapers and government documents about food policy. I returned to Ghana for additional archival research during my PhD course at SOAS, University of London where I extended my research into the culinary history of Ghana. I am most grateful to the staff of the search room at PRAAD in Accra, as well as the staff in Kumasi, Cape Coast, and Koforidua. I collected additional archival materials on Ghanaian proverbs and myths, as well as Ghanaian archaeobotany upon my return from Ghana at SOAS, University of London, the British Library, and the Bodleian Library in Oxford.

The process of writing the book was a series of stops and starts and reformulations of the approach I wanted to take. It took some time to learn how to write from the Ghanaian perspective when much of the material was written from the Danish, Dutch, and British perspectives. My tireless supervisor, John Parker was instrumental in this process. He aided greatly in helping to pilot the plane and to bring it in for a landing.

Abingdon, UK — Brandi Simpson Miller

Acknowledgments

With sincere thanks to John Parker, my supervisor, and thanks to the kind staff of the search rooms at PRAAD, Accra, Cape Coast, and Kumasi, at the SOAS library in London, and the Bodleian Library in the David Reading Room in Oxford. Many thanks to The Friends of Abingdon Abbey Buildings Trust, Catherine Walter—Linacre College, Oxford, Shehzad Naqvi—Linacre College, Oxford, and Clive and Kerstin Bowsher of Abbey Baptist in Abingdon for their support, advice, and patience. Thank you to everyone who gave their time and thoughts in interviews and consultations, and to all the people who helped in many other ways over the course of completing this thesis. It would never have been possible without your help. I gratefully acknowledge the sacrifices of my mother Nancy Merilan Leath Simpson and of the man that raised me, Jimmy Lee Simpson. I also recognise the sacrifices of my children—Henry Joseph Anderson III and Caira Isabella Anderson. A special thank you to my husband and partner, Tim Miller without whose hard work and unfailing support this work would not have been possible.

Praise for *Food and Identity in Nineteenth and Twentieth Century Ghana: Food Fights and Regionalism*

"This book presents a fascinating account of the complex regional foodways, and the different eco-culinary zones of The Gold Coast/Ghana from the nineteenth to the twenty-first century. It uses an impressive range of primary and other sources and the approach taken by the author would provide an excellent model for exploring the history of culinary culture of other African nations. The book also provides a particularly insightful investigation into the history of the gendered division of labour, and the crucial role played by African women in growing, preparing, selling and cooking food"

—Dr Igor Cusack, *UK*

"This book on the social history of food and cooking in Ghana represents a major interdisciplinary accomplishment. She has, like an expert quilter, carefully and critically pieced together her own and others' research on material culture, history, geography, sociology, politics, food studies, technology, gendered relationships, theologies, and more, to create a timeless and fascinating work of beauty and great value"

—Dr. Fran Osseo-Asare, PhD, Independent Scholar, *USA*

Contents

List of Figures

1

In Search of Ghanaian Food

On my first visit to Ghana, it became apparent to me that there was plenty to eat. Yet the standard way of thinking about food in Africa is that there is very little. I was accompanying Adelaide, the cook from the guest house I was staying in, to the Adomo market. She thought this relatively small market would be a good introduction to food shopping in Accra. We made our way through narrow, winding paths that cut through what seemed like hundreds of stalls of smoked fish, enormous yams, garden eggs (small yellow eggplant), and many other items of produce. Porters at the Adomo market helped carry our purchases to the taxi stand for three Ghana cedi. Accra boasted a wide variety of restaurants, and food was available for sale on many street corners and in each market. Whilst travelling through Ghana, one could purchase banku (fermented maize dumpling) and fried tilapia at rest stops, and on the roads at each speed bump numerous food vendors congregated, lifting their offerings up to the bus windows (Fig . 1.1).

The accepted Western way of thinking about food scarcity in Africa is reflected in Western popular culture. In the BBC show 'The Misadventures of Romesh Ranganathan,' which aired in November of 2019, comedian Romesh Ranganathan travelled to Ethiopia. His concern in this episode

B. Simpson Miller, *Food and Identity in Nineteenth and Twentieth Century Ghana*,
Food and Identity in a Globalising World,
https://doi.org/10.1007/978-3-030-88403-1_1

Fig. 1.1 Banku and Tilapia in Accra

was whether he would "find any food", and his mission was to ascertain whether Ethiopia's reputation as "really a food-starved dust bowl" was justified.[1] Ranganathan cited the Live Aid concert of 1985 as responsible for setting his expectations with regards to food—or the lack thereof—in Africa. Likewise, adverts from organisations like Save the Children and World Vision often feature black faces and are designed to convince Westerners that children are starving all over Africa.

This project attempts to counteract the common view of African foodways as being characterised by scarcity by examining the cooking and eating practices of the Gold Coast (present-day Ghana) from the beginning of the nineteenth century to 2014.[2] The work asks what structured the available choices of foods, how food was cooked and eaten, and how these choices have changed over the last two centuries. It seeks to understand how cooking, eating, and identity were and still are connected to the local microclimates and the resulting ecologies in each of Ghana's

[1] 'The Misadventures of Romesh Ranganathan, Series 1, Episode 2', BBC, accessed June 8, 2020, https://www.bbc.co.uk/programmes/b0bb3gnc.

[2] Here and throughout the thesis, unless otherwise indicated I use the terms for Gold Coast and Ghana anachronistically and interchangeably.

three major eco-culinary zones. My argument is that the foodways of Ghana have been resistant to change over centuries of interaction and trade with the rest of the world, and this project seeks to explain the reasons for the resilience of Ghanaian foodways.

This work is not an exploration of how the African American Diaspora relates to West Africa and the tensions between its active creativity in survival—the central issue of African American culture and a central debate. Nor is this project an exploration of the African Diaspora generally. These topics have been addressed in numerous ways by other scholars.[3] This work will approach the history of cooking and eating in Ghana by making visible the 'consuming geographies' of Ghana—the geographical specialisations that connect with a specific taste and place.[4] Some other themes treated in the project necessarily include a consideration of how food is linked to cultural ideas of masculinity and femininity in Ghana, which, like Ghana's foodways, appear resistant to change. This project will also appraise the ways in which commensality has been a daily expression of values. It investigates the ways in which food—the foundation of every economy—was used as the central pawn of political strategies, where food expressed power. This work also marks how food sharing created solidarity and delineated boundaries, and how men and women in Ghana have used food to define themselves differently from their neighbours.

Using an interdisciplinary approach, this project charts food practices by using traditional historical archival materials analysis, cookbooks used as historical texts, and materials from other fields, such as anthropology, sociology, linguistics, and crop science. This interdisciplinary approach allows for the illustration of a range of symbolic, material, and ecological perspectives that can be used to understand food selection and its

---

[3] Jessica Harris, *High on the Hog: A Culinary Journey from Africa to America* (New York: Bloomsbury USA, 2011); Judith A. Carney, *Black Rice: The African Origins of Rice Cultivation in the Americas*, New Ed edition (Cambridge, Mass.: Harvard University Press, 2002); Jessica B. Harris, *The Welcome Table: African-American Heritage Cooking* (New York: Simon & Schuster, 1995); Judith Carney, *In the Shadow of Slavery: Africa's Botanical Legacy in the Atlantic World* (Berkeley: University of California Press, 2009); Sidney Mintz, *Tasting Food, Tasting Freedom: Excursions into Eating, Power, and the Past* (Boston: Beacon Press, 1997).

[4] David Bell and Gill Valentine, *Consuming Geographies: We Are Where We Eat* (London: Routledge, 1997).

associated historical consequences in Ghana. Commenting on the tendency of scholarship to focus upon hardship, James McCann stated that there is another path toward the study of cooking in Africa: "The second approach, the one I survey here, is food as a creative composition at the heart of all cultural expressions of ourselves as humans, that is to say, cuisine".[5] In writing about recipes and cooking as historical knowledge, this work effectively contributes to a new understanding of the social history of Ghana and the agency and humanity of its people. In so doing, this work augments knowledge in the fields of both history and food studies.

Food studies is the study of the relationships between food and the human experience, examined from a variety of perspectives and from a range of places in the food system, from production to consumption. It is typically interdisciplinary and investigates people's relationships with food from a range of humanities and social science perspectives.[6] However, when it comes to African food studies, food security remains the dominant theme over and above the study of the relationships between food and the human experience.[7] Reay Tannahill, in her classic book *Food in History* (1973), remarked that food in Africa was not consistently plentiful enough to be examined for its skill and technique: "When shortages are a part of everyday life, filling the stomach is the only art."[8] Early food

[5] James C. McCann, 'Writing on the African Pot: Recipes and Cooking as Historical Knowledge', in *Writing Food History: A Global Perspective*, eds. Kyri W. Claflin and Peter Scholliers (New York: Berg, 2012), 200.

[6] Jeff Miller, *Food Studies: An Introduction to Research Methods* (Oxford: Berg, 2009), 3.

[7] Jérôme Destombes, 'Nutrition and Chronic Deprivation in the West African Savanna: North Eastern Ghana, c. 1930–2000' (unpublished PhD thesis, University of London, 2001); Jérôme Destombes, 'From Long-Term Patterns of Seasonal Hunger to Changing Experiences of Everyday Poverty: Northeastern Ghana c. 1930–2000', *Journal of African History* 47, no. 2 (2006): 181–205; Roger M. Blench, *Natural Resource Management in Ghana and its Socio-Economic Context* (London: Overseas Development Institute, 1999); Paul Shaffer, 'Seasonal Hunger in the Northern Territories of the Gold Coast, 1900–40', *Economic History of Developing Regions* 32, no. 3 (September 2, 2017): 270–300; Gareth Austin, Joerg Baten, and Bas Van Leeuwen, 'The Biological Standard of Living in Early Nineteenth-Century West Africa', *Economic History Review* 65, no. 4 (2012): 1280–1302; Amanda L. Logan, '"Why Can't People Feed Themselves?": Archaeology as Alternative Archive of Food Security in Banda, Ghana', *American Anthropologist* 118, no. 3 (2016): 508–24; F.M. Purcell, 'Nutrition In The Colonial Empire', *British Medical Journal* 2, no. 4100 (1939): 294–96; A. W. Cardinall, *The Natives of the Northern Territories of the Gold Coast; Their Customs, Religion and Folklore* (London: G. Routledge, 1920).

[8] Reay Tannahill, *Food in History* (New York: Stein and Day, 1973), 317.

studies approaches to African foodways refused to begin to assess the skill and technique of African cooks based upon the assertion that there was not enough food in Africa to study. This assertion is a myth that needs to be challenged.

Equally prevalent in the scholarship were examinations of nutrition and the effects of colonialism and industrially produced foreign foods on the health of populations.[9] This scholarly focus on scarcity and the preoccupation with colonial legacies obscured the understanding of cuisine in Africa as an historical process. When I began this project, my research interests were focused upon how imported foods were being incorporated into daily cooking and eating, and the social effects of the adoption of those imported foods. However, it became clear that there was no adequate foundational historical account of Ghanaian food with which to compare contemporary foodways. This stands in contrast to studies of Western countries, which have volumes dedicated to historical food studies.[10]

Some of these volumes of "global" historical food consumption such as Flandrin and Montanari (1999), Kiple and Orrelas (2000), and Davidson (2014) are remarkably thin on analysis of West African food.[11] One exception is the work of Kiple and Orrelas, who collapsed the entire region of West Africa from southern Mali to the coast into a comparatively small section, and then focused on scarcity and malnutrition. Perhaps approaches were coloured by the view that Africans remained in need of rescuing from the cultural corruption of colonialism or the fickle

[9] Deborah Johnston et al., 'Review: Time Use as an Explanation for the Agri-Nutrition Disconnect: Evidence from Rural Areas in Low and Middle-Income Countries', *Food Policy* 76 (2018): 8–18; Dr. K.A. Busia, 'Sekondi-Takoradi Social Survey—School Feeding', 9 December 1947, ADM 23/1/2768, PRAAD Accra.

[10] Ken Albala, *The Food History Reader* (London: Bloomsbury Academic, 2014); Margaret L. Arnott, *Gastronomy: The Anthropology of Food and Food Habits* (Netherlands: Walter de Gruyter, 2011); Charles Camp, *American Foodways: What, When, Why and How We Eat in America* (Little Rock: August House, 1989); Kenneth F. Kiple and Kriemhild Coneè Ornelas, eds., *The Cambridge World History of Food* (Cambridge: Cambridge University Press, 2000); Jeffrey M. Pilcher, ed., *The Oxford Handbook of Food History* (New York: Oxford University Press USA, 2012).

[11] Jean-Louis Flandrin and Massimo Montanari, eds., *Food: A Culinary History from Antiquity to the Present*, trans. Albert Sonnenfeld (New York: Columbia University Press, 1999); Kiple and Ornelas, *The Cambridge World History of Food*; Alan Davidson, *The Oxford Companion to Food* (Oxford: Oxford University Press, 1999).

whims of a changeable environment.[12] Maybe the problem was that Africans were thought to have no agency and were dominated by colonial cultural influences, so there were thought to be no local cuisines to assess and thus, no reason to investigate consumption.[13] Even when the scholarship discusses new world foods and their impact on the old world, Africa was frequently absented from discussions.

Scholars of the Columbian exchange from the sixteenth to the eighteenth century provided this project with a starting point. The work of scholars like Alfred Crosby and his monograph *The Columbian Exchange* led me to ask questions about the effects of new world crops—such as maize—upon the old world of Africa.[14] In his chapter entitled "New World Foods and Old-World Demography", Crosby gave an overview of how new world foods maize and cassava, provided higher yields for African farmers, resulting in significant population growth in the nineteenth century. In addition to providing higher yields, cassava was drought and insect resistant. Both maize and cassava proved to be successful competitors with native millet and sorghum on the African continent. Crosby's work first piqued my curiosity about Africa's contributions to the broad fields of agricultural history and ecology.

A geographer of West Africa and the African Diaspora, Judith Carney built upon the work of Crosby in her chrono-thematic approach to single commodities food studies in *Black Rice: The African Origins of Rice Cultivation in America*.[15] Utilizing travel narratives, plantation records, patent requests, slave ship logbooks, and inventors accounts, she examined the effects of the transfer of rice growing technology from the Guinea highlands to South Carolina and other parts of the Americas. A geographical full circle narrative that touched upon themes of gender and labour during the transatlantic slave trade, this investigation focused on

[12] Jean-Pierre Chretien, 'The Historical Dimension of Alimentary Practices in Africa', *Diogenes* 36, no. 144 (December 1988): 92–115; Elias Mandala, 'Beyond the "Crisis" in African Food Studies', *Journal of The Historical Society* 3, no. 3–4 (2003): 281–301.

[13] Fran Osseo-Asare, *Food Culture in Sub-Saharan Africa* (Westport: Greenwood Press, 2005).

[14] Alfred Crosby, The Columbian Exchange: Biological and Cultural Consequences of 1492 (Westport: Greenwood Publishing Company, 1972).

[15] Judith A. Carney, *Black Rice: The African Origins of Rice Cultivation in the Americas* (Cambridge: Harvard University Press, 2002).

the production and processing of rice and its connection to West African gendered divisions of labour. Carney's work converged upon the commercial transfer of food knowledge, processing and custom from Africa to the Americas, and it served as a guide for how to investigate the connection between political-economic and symbolic value creation, as well as the social construction of memory through food.

The botanic and faunic legacy of Africa in the Atlantic World is the topic of Judith Carney and Nicholas Rosomoff's *In the Shadow of Slavery*.[16] The main focus of this study is on how subsistence was negotiated during the trans-Atlantic slave trade. Using an interdisciplinary approach which included travel narratives, visual evidence and fieldwork, this monograph considered how the biological knowledge transfer of subsistence crops from the old world of Africa to the new world of the Americas was critical to the slave trade and the development of the Atlantic world. This work was wide in scope, encompassing the Americas, the Caribbean, and Africa, and investigated the flow of foods from Africa to the Americas. The works focus on subsistence crops and its examination of the symbolic meaning of plants and foodways in shaping identities in the African Diaspora prompted me to consider similar themes in this study.

Another Atlantic World offering that surveyed the connection between African foodways and the Americas is Jessica Harris' *High on the Hog: A Culinary Journey From Africa to America*.[17] Harris aimed to showcase the agency of Africans in shaping regional cuisines in the new world via selection, knowledge, cultivation, and preparation of African foods during the era of the slave trade and beyond. Like Carney and Rosomoff, Harris examines the provisioning of forced labourers as a lens through which to conduct her study. Although her focus was the African influence on North American foodways, in the process Harris explored the foodways from North and West Africa. She situated cultural adaptations within the context of globalization and diffusion that have been extant for centuries. *High on the Hog* was part of the contemporary trend of migration and

---

[16] Judith Carney and Richard Rosomoff, *In the Shadow of Slavery Africa's Botanical Legacy in the Atlantic World* (Berkeley: University of California Press, 2009).

[17] Jessica Harris, High on the Hog: A Culinary Journey from Africa to America (New York: Bloomsbury, 2011).

agency in the study of foodways and has helped me to consider themes of globalization and migration in my own approach to this project.

An important contribution to African scholarship in Atlantic World food history is J. D. La Fleur's *Fusion Foodways of Africa's Gold Coast in the Atlantic Era.*[18] Responding to calls for Africa to be integrated into Atlantic history, La Fleur applied Crosby's wide-angle approach to the Columbian exchange on the Gold Coast from the fifteenth to the eighteenth centuries. Focusing on the adaption of plantain, maize, and cassava as key staples on the Gold Coast, he examined the processes farmers underwent in choosing new world food crops and considered the cultural symbolism and health consequences related to those choices. Recognizing the creativity and ingenuity with which Ghanaians engaged with the opportunities and challenges of the Atlantic slave trade, *Fusion Foodways* placed starches at the center of its narrative. This project opens in the nineteenth century, where La Fleur's investigation ends. Carrying on from *Fusion Foodways*' study of what Africans grew and ate in opening centuries of Atlantic commerce, it explores the ways in which New World starches were transformed to make them acceptable to local characterisations of what was considered the proper meal. It also asks how the Gold Coast—being a site of converging foodways constituted by locals, Europeans, and a wider African diaspora—retained concepts about the proper meal despite their participation in the wider Atlantic world economy in the last two centuries. In a further extension of La Fleur's work on the choosing of starches, the processing techniques, cooking methods, and eating customs of starchy staples in Ghanaian climates take centre stage in this work.

Historiographical Context

Close scrutiny of the food studies subfield as it related specifically to Africa revealed a highly specialized body of work that gives us some idea about what was being eaten in Africa. Sidney Mintz and Christine Du Bois' typology of food studies is most useful for sketching the history of

[18] J. D. La Fleur, *Fusion Foodways of Africa's Gold Coast in the Atlantic Era* (Leiden: Brill, 2012).

the subfield and for tracing major trends in the study of African foodways.[19] Mintz and Du Bois identified seven subsections in the field of food studies: classic food ethnographies, single commodities, food and social change, food insecurity, eating and ritual, eating and identities, and instructional materials. Much of the research on African foodways predates Mintz and Du Bois' typology. Nevertheless, the typology is still useful for understanding the subfields of food studies in Ghana in particular, in Africa generally, and in the greater Atlantic World and its African diaspora.

Food studies of Ghana originated with anthropologists like R. S. Rattray, and Meyer and Sonia Fortes in the early twentieth century.[20] Their classic ethnographical works were commissioned as part of colonial information seeking initiatives about the general material culture of different ethnic groups under their administration. A notable exception to this trend was the work of anthropologist Margaret Field. Her 1931 pamphlet "General Survey of Gold Coast Food", surveyed the foods prepared every day in the geographical ecologies of the coast and in the forested centre.[21] Collected directly from locals and from direct observations of food preparation, the foods and recipes are organised in categories such as starchy foods, protein foods, and oil seeds and nuts. Intended as instructional material, very little of the social, historical, or political meanings of cooking and eating were explored in this brief work. However, Field's work provided comprehensive information about more staple foods from this time period and their preparations than any work of its time.

Addressing Ghanaian foodways specifically was Alice Dede's cookbook *Ghanaian Favourite Dishes* (1969).[22] Printed twelve years after independence, Dede's book had as its goal to collect recipes "from all over the country and present a complete diet for the whole nation". Her collection

---

[19] Sidney W. Mintz and Christine M. Du Bois, 'The Anthropology of Food and Eating', *Annual Review of Anthropology* 31 (2002): 99–119.

[20] R. S. Rattray, *The Tribes of the Ashanti Hinterland* (Oxford: Clarendon Press, 1932); M. Fortes and S. L. Fortes, 'Food in the Domestic Economy of the Tallensi', *Africa* 9, no. 2 (1936): 237–76; Cardinall, *Natives*; A. W. Cardinall, *The Gold Coast, 1931* (Government Printer: Accra, 1932).

[21] M. J. Field, 'Gold Coast Food', *Petits propos culinaires* 43 (1993), [*Gold Coast Food* (Achimota: Achimota College Press, orig. 1931)].

[22] Alice Dede, *Ghanaian Favourite Dishes* (Accra: Anowuo Educational Publications, 1969).

falls under the subsections of instructional materials in the Mintz/Du Bois typology. This collection reflects many of the foodways of the time and, as this was a cookbook, it contained no historical analysis of the ingredients or preparation methods contained therein. An expansion of Field's 'Survey', this cookbook was intended to reflect the new national pride of Ghana and is a wonderful resource in its time on the preparation of foods from each eco-culinary zone of Ghana.

Jack Goody's classic *Cooking, Cuisine, and Class* (1982) was part of a long nineteenth- and twentieth-century anthropological trend towards the studying of less-industrialised food production systems.[23] Goody took a comparative approach toward answering the central question of why a differentiated "haute cuisine" had not emerged in West African nation states like Ghana. Goody did not approach the issue of foodways through cooking, but through production. His classic comparative ethnological work was meant to demonstrate that there was a connection between the technology used to produce food and food knowledge—such as a hand axe or an ox plough and oral transmissions versus the written word—and the development of elaborate courtly cooking cultures in nations like Ethiopia. Using these parameters, Goody concluded that Ghana did not have a cuisine.

Goody's conclusion produced (and is still producing) a response that has been elaborated upon by scholars like Gracia Clark in her book *Onions Are My Husband* (1994). Clark's ethnographical work was concerned with food and social change. With a focus on the labour of cooking, she sought to uncover what cooking meant in terms of social relationships. Clark took a gendered approach, as she detailed the lengthy processing methods of staple starches and examined the vital role women had as the reproducers of identity through their cooking. Both *Onions*, and her more recent work "From Fasting to Fast Food in Kumasi"—part of a collection of essays in honour of Jack Goody—focused upon who was doing the preparation and what the social implications of this labour were.[24]

---

[23] Jack Goody, *Cooking, Cuisine and Class: A Study in Comparative Sociology* (Cambridge: Cambridge University Press, 1982).

[24] Gracia Clark, *Onions Are My Husband: Survival and Accumulation by West African Market Women* (Chicago: University of Chicago Press, 1994); and 'From Fasting to Fast Food in Kumasi, Ghana',

Clark's use of food in her analysis has shed much light on social processes in Ghana. Fran Osseo-Asare, an internationally recognized authority on African food and culture, has expanded upon Clark's approach by deeply engaging with Ghanaian food preparation. Her published works—*A Good Soup Attracts Chairs* (1993), *Food Culture in Sub-Saharan Africa* (2005), and *The Ghana Cookbook* (2015)—further the discussion of food and social change in West Africa.[25] In her Ghanaian recipe books and on her Betumi website and blog, sociologist Osseo-Asare has worked to popularize Ghanaian cooking and the adaptation of its traditional cooking methods to modern kitchen spaces.[26] Her body of work has brought about the knowledge of African food in America and is a valuable resource for students of West African social history.

In the twentieth century, the study of the broader geography of single commodities and substances and their impact on global consumption came about beginning with Sidney Mintz's 1985 *Sweetness and Power*.[27] James McCann continued this trend with his study of maize with *Maize and Grace: Africa's Encounter with a New World Crop, 1500–2000* (2005).[28] In this monograph McCann chronicled the spread of different varieties of the crop throughout Africa and related how it became a principal staple. Although Ghana—specifically the Asante and their relationship with maize production and its political implications—was surveyed in this work, this monograph is a comparative work intended as a cautionary tale about the dangers of the monocropping of hybrid maize. McCann did, however, acquaint the food studies scholar with the dominant varieties of maize—flint and floury—that complemented existing African crops sorghum, millet, and yam. Maize, a crop that could be integrated

in *Food Consumption in Global Perspective: Essays in the Anthropology of Food in Honour of Jack Goody*, eds. Jakob Klein and Anne Murcott (Basingstoke: Palgrave Macmillan, 2014).

[25] Fran Osseo-Asare, *A Good Soup Attracts Chairs: A First African Cookbook for American Kids* (Gretna: Pelican Publishing Company, 1993); Fran Osseo-Asare and Barbara Baëta, *The Ghana Cookbook* (New York: Hippocrene Books, 2015); Osseo-Asare, *Food Culture in Sub-Saharan Africa* (Westport: Greenwood Press, 2005), xiii.

[26] BETUMI: African Cuisine, accessed 31 July 2018, http://www.betumi.com/home.

[27] Sidney Wilfred Mintz, *Sweetness and Power: The Place of Sugar in Modern History* (New York: Penguin Books, 1986).

[28] James McCann, Maize and Grace: Africa's Encounter with a New World Crop, 1500–2000 (Cambridge: Harvard University Press, 2005).

into the existing system of intercropping, boosted nutritional capacity and reinvented Africa's food supply over the last 500 years since its introduction. Being a continent-wide study of maize, *Maize and Grace* was necessarily broad in scope and focused upon a single micro-climate in Ghana, comparing it with other microclimates in different regions of Africa. This work is an expansion of McCann's work in that all three of the major climates of Ghana and their relationship with maize will be explored.

Continuing with the trend of single commodity and substance studies, Emmanuel Akyeampong's work *Drink, Power, and Cultural Change* (1996) was an important first for Ghana. Akeyampong delved into the interrelationship between social and ceremonial consumption of alcohol and the exercise of power in the period from 1860 to 1900. In his book, gender roles—in terms of traditional male dominance exercised through the consumption of alcohol—as well as the changing fortunes of women were central leitmotifs of this book.[29] Although technically not concerned with food and eating, the work showed just how much could be learned about social change in the transition from the slave trade to the legitimate trade in Ghana by carefully exploring the history of drink. Akeyampong's monograph fell under three categories in Mintz and Du Bois' typology: the study of single commodities and substances, social change, and drink as it relates to ritual. Colonialism and commercial revolution in items like rubber and cocoa, spurred political, social, gender, economic, and religious changes. Akyeampong examined these upheavals through the procurement, trade for, and consumption of alcohol. In his approach to alcohol, Akyeampong drew in the Akan in the forest zone, the Ga-Adangme on the coast, and the Ewe in Togoland with greater depth than previously explored in Ghanaian history. Akyeampong's approach to the study of alcohol in Ghana provided this project with a framework for the study of the differences in what people ate, where they ate, how they ate, and the changing social significance of cooking and eating in terms of class development in the early colonial period.

---

[29] Emmanuel Akyeampong, Drink, Power, and Cultural Change: A Social History of Alcohol in Ghana, C. 1800 to Recent Times (Portsmouth: Heinemann, 1996).

Turning to African food studies generally, Audrey Richards' careful scholarship led her to being regarded as the founder of nutritional anthropology. In line with the tradition of the classic anthropological ethnography, Audrey Richards' monograph *Hunger and work in a savage tribe: a functional study of nutrition among the Southern Bantu* set out to reveal how the fundamental urge for food shaped human institutions in southern African societies.[30] Richards explored Bemba production and the influence of its extreme seasonality on local institutions in *Land, Labour, and Diet in Northern Rhodesia*.[31] Much like Meyer Fortes among the Tallensi of Ghana, her body of work was associated with a single ethnic group in Africa, the Bemba. Typical of ethnographies of the time, Richards' functional study of the economics of the Bemba sought to demonstrate the consequences of the introduction of a money economy, taxation, and migration to the Bemba lifestyle. Relying on British colonial ethnographic studies and extensive field work, she detailed the foodways, including an analysis of the 'Perfect Meal', and brought to light the cultural weightiness of the staple starch, finger millet. She also investigated the uses European imports were put to such as bread, sugar, and sardines—which eased the preparation of the main meal. Although technically an anthropological study of economy and food insecurity among the Bemba, Richards analysis of the perfect meal was an important guide toward my own understanding of the relationship between eating and identities in the Gold Coast.

A social and economic historian of Nigeria, Eno Blankson Ikpe published a political economy of food entitled *Food and Society in Nigeria* in 1994.[32] This work linked Nigeria's foodways to its progressive integration into the world economy. Considered by Sidney Mintz and Christine Du Bois in their typology of food studies to be a classic ethnography, it was also considered to be a study of food and social change and food insecurity. Her main argument—connected to the Biafran war of 1966–1970—was

[30] Audrey Richards, *Hunger and Work in a Savage Tribe: A Functional Study of Nutrition among the Southern Bantu* (London: George Routledge & Sons, 1932).

[31] Audrey Richards, *Land, Labour and Diet in Northern Rhodesia: An Economic Study of the Bemba Tribe* (Oxford: Oxford University Press, 1939).

[32] Eno Blankson Ikpe, *Food and Society in Nigeria: A History of Food Customs, Food Economy and Cultural Change, 1900–1989* (Stuttgart: Franz Steiner, 1994).

that large-scale relief efforts integrated Nigeria into the international food market. Through this conflict the administration learned that subsisting on imported foods increased their political power, even if they had to borrow hard currency to do so. Ikpe's work investigated the amalgamation of imports into daily foodways and how the changing economic landscape contributed to these changes. Limited to the twentieth century, her work placed war at the centre of her investigations as an agent of dietary change and did little to address conceptualizations of identity through food. Her periodization of consumption patterns—precolonial, colonial, Civil War, oil boom, structural adjustment—and investigation into how imports were integrated into the foodways of Nigeria, provided my own work with a guide for approaching similar questions in Ghana.

Jeremy Rich has advanced the field of African urban food history with his detailed case study of Libreville in *A Workman is Worthy of His Meat* (2007).[33] An analysis of interdependent food relationships between Libreville and the French empire during the period 1840–1960, this work was at once a classic food ethnography, a study of food and social change, an examination of food insecurity in Libreville, and an analysis of food and its role in the construction of identities. Making use of missionary and official sources, as well as fieldwork conducted in Libreville, Rich focused his query on the change over time in food supply patterns and social relationships. Generally describing local staples and available imports, Rich eschewed a more detailed consideration of how the food was prepared in favor of broad descriptions of consumption patterns and how they changed over time. His approach to periodization and ecology, as well as his careful analysis of the social hierarchy constructed around food remuneration of locals by Europeans, informed the periodization and focus of this project.

Like Jeremy Rich, James McCann's study of the historical development of the African landscape in *Green Land, Brown Land, Black Land* also made use of case studies to challenge the narrative of ecological crisis in Africa.[34] He argued that Africa's landscape dynamism was not caused

[33] Jeremy Rich, A Workman Is Worthy of His Meat: Food and Colonialism in the Gabon Estuary (Lincoln: University of Nebraska Press, 2007).

[34] James McCann, Green Land, Brown Land, Black Land: An Environmental History of Africa, 1800–1990 (Portsmouth: Heinemann, 1999).

by human degradation, but that landscapes were the product of human action. Covering the period from 1800 to 1990, this monograph was comparative in its approach, including a chapter on biodiversity in Ghana's Upper Guinea forests. McCann asked questions about how environmental history shaped geo-political outcomes in terms of crop availability food production systems. McCann's work in *Green Land* initiated a closer examination of food and its wider implications for Ghana's future rural landscape. *Green Land* was broad in scope in that McCann compared seven different countries in Africa. Further, when it came to Ghana, he limited his case study to the Asante in the forested centre and that empire's relationship to maize and the forest fallow system—ideas first put forth by Ivor Wilks in the 1990s. *Green Land's* study of food and social change contributed to a better understanding of how the ebb and flow of commercial agriculture (cocoa especially) in the late nineteenth to the early twentieth centuries produced an increased reliance on maize all over Ghana due to its adaptability, low labour inputs and high yields.

Part of the trend of the study of migration and food in Africa, McCann took a *longue durée* approach toward establishing how trade and migration influenced the ingredients and technology of African cookery since well before the sixteenth century. In his foray into the foodways of West Africa, James McCann controverted Goody's charge that West Africa had no cuisine in his book *Stirring the Pot* (2009).[35] A history of food, ecology, and agriculture of the African continent by means of recipe analysis, McCann confirmed the agency African people as exhibited in food choices and processing technologies. *Stirring the Pot* is a broad geographical survey, covering Ethiopia, Ghana, Brazil, and the Caribbean from 1500 to 2000. Thus, it fails to closely compare localized foodways from different microclimates, preferring instead to investigate the commonalities of the cuisines of West Africa. However, McCann's in-depth examination of common staple ingredients and cooking techniques helped to support his view of cuisine as an expression of cultural identity and food as a living history of cultural change. McCann's work informs this project

[35] James McCann, *Stirring the Pot: A History of African Cuisine* (Athens: Ohio University Press, 2009).

with his caution to resist fixed notions of "authentic" or "traditional" food and in his urging to adopt themes of migration, hybridity and flow.

An ethnoarchaeologist with a focus on food security, Amanda Logan also explored the impact of cooking techniques on African foodways over time. Her thesis 'A History of Food Without History' relied upon anthropological and archaeobotanical data, traditional historical sources, and oral history to understand what the Banda people of Ghana ate from 1000 AD to the present.[36] At once an ethnography and a study of food and social change, Logan argued that there was not "a permanent or fixed shortfall in African agricultural systems and village life" that may have led to an indiscriminate wholesale adoption of maize and cassava when it was introduced in the sixteenth century.[37] In this way Logan's work echoed similar findings to La Fleur's *Fusion Foodways*. Logan went on assert that "The unstable conditions that led to the adoption of maize and cassava thus occurred rather late in time, in the late 19th and early 20th centuries." She considered questions about how identity was transmitted through bodily practices which were resistant to change. Her work has great historical depth but was limited to the study of the Banda district and changes in their food security. Nevertheless, by observing the deep history of what people in this microclimate ate and how that changed over time, Logan provided my own study with an understanding for how culture was made and remade with food. My own geographical scope is broader, covering three major eco-culinary zones and several ethnic groups. Additionally, my project is limited to the period of economic change from the slave trade to colonization, up to the present day. An extension of Logan's investigations into the connection between food and memory, this work investigates how newer foods like maize and cassava were slotted into existing food systems during the pre- and post-colonial period.

To date, much of the research since Goody's classic work about the Gold Coast has generally revolved around concerns over food security and food and social change in the twentieth and twenty-first century.

---

[36] Amanda Logan, 'A History of Food Without History: Food, Trade, and Environment in West-Central Ghana in the Second Millennium AD' (PhD diss., University of Michigan, 2012).

[37] Ibid., 311.

Several studies have been valuable in identifying general patterns of the creation of (or indeed, resistance to the creation of) national cuisines in African countries.[38] What might we learn from fine-grained and in-depth accounts of cooking and eating in African countries like Ghana as part of the food studies academic subfield in all its disciplinary, organisational and intellectual complexity? What would in-depth analysis of the daily eating habits of people in the different microclimates of Ghana allow us to see that the large-scale studies or single focus ethnographies do not? An examination of food and social change, and food and identities in Ghana, this project is intended to be an investigation of the often-neglected period of food history during the transition away from the slave trade. It concludes with an exploration of food, identity, and nationalism in modern Ghana—of which Richard Wilk's *Home Cooking in the Global Village* and Jeffrey Pilcher's *Que Vivan Los Tamales* are prominent examples.[39]

Sources

As part of my Masters on food and nationalism in post-independence Ghana, and again as part of my doctorate on the history of pre, colonial, and post-independence foodways, I was fortunate enough to undertake archival trips to Ghana in the summer of 2014 and again in the autumn of 2016. Before my initial departure I consulted David P. Henige, and R. E. Dumett's surveys of the holdings of the Public Records Administration and Archives Department (PRAAD) to familiarize myself with the available government records, regional and district holdings,

[38] Jack Goody and Esther N. Goody, 'Food and Identities: Changing Patterns of Consumption in Ghana', *Cambridge Anthropology* 18, no. 3 (1995): 1–14; Igor Cusack, 'African Cuisines: Recipes for Nation-Building?', *Journal of African Cultural Studies* 13, no. 2 (2000): 207–25; Paul Nugent, 'Do Nations Have Stomachs?: Food, Drink and Imagined Community in Africa', *Africa Spectrum* 45, no. 3 (2011): 87–113; Logan, 'A History of Food'; Logan, '"Why Can't People Feed Themselves?"'; Brandi Simpson Miller, 'Food and Nationalism in an Independent Ghana', in *The Emergence of National Food*, ed. Atsuko Ichijo, Venetia Johannes, and Ronald Ranta (London: Bloomsbury Academic, 2019).

[39] Richard Wilk, *Home Cooking in the Global Village: Caribbean Food from Buccaneers to Ecotourists* (Oxford: Berg, 2006); Jeffrey M. Pilcher, *Que Vivan Los Tamales!: Food and the Making of Mexican Identity* (Albuquerque: University of New Mexico Press, 1998).

and newspaper collections at the main branch in Accra.[40] From these publications I learned that Secretary for Native Affairs (ADM) series would be the place to start for information, especially about the Asante, the Fante, and the Gã populations, as well as information about the Northern Territories. Whilst I visited the PRAADs in the regional archives of Kumasi, Koforidua, and Cape Coast, I conducted the majority of my archival research in the Accra PRAAD. In addition to documents on the period following the Ghanaian independence from Britain in 1957, the Accra PRAAD contains the largest body of documents and holds a rich collection of sources from the colonial era. The regional archives provided an important pool of complementary sources for my research on Ghanaian food history.

In addition to the articles I consulted before my archival visits, my supervisor, John Parker, was most helpful in advising me about the District Record Books in the ADM series, valuable for their anthropological documentation of different ethnic groups. (In the early colonial period, oral histories were collected in District Record Books as part of the Crown Lands Bill of 1897, meant to ascertain how tribes came to acquire the lands they currently inhabited.) On my second trip I also consulted with Kojo Bright, Records Assistant at PRAAD Accra who advised me to consult the CSO series for mentions of food, and the ADM series for information on the colonial food supply and food economy of the Northern Territories. The ADM series was particularly helpful, providing detailed information about the food supply, costs, and imported foods from 1909 to 1977. A systematic consultation of the indexes and the resultant review of the ADM series also yielded documents on different food related industries (like the fishing industry) and any issues or information related to food supplies. The CSO series in particular yielded much information about the 1930s local production of imported foodstuffs (import substitutions), food supplies to the Northern Territories, and Annual General Reports.

[40] David P. Henige, 'The National Archives of Ghana: A Synopsis of Holdings', *The International Journal of African Historical Studies* 6, no. 3 (1973): 475–86; Raymond E Dumett, *Survey of research materials in the National Archives of Ghana* (Gemund: Lempp Verlag, 1974).

Together with extensive colonial records, the PRAAD in Accra archive yielded many newspapers from the colonial period beginning in the 1940s, to the period immediately after independence in 1957. On my first visit, I systematically searched newspaper articles and advertisements for any mention of food and photographed and catalogued my findings. The holdings of PRAAD Cape Coast, Kumasi, and Koforidua consisted mainly of regional government documents which I examined, looking for any mention of food, markets, or food supplies. In these regional archives I reviewed the same ARG and ADM series as I judged these to be my best source of information on anything food related.

In Kumasi, I was advised by Harry Ekuful, Records Assistant, to view the ARG series which contained, census reports, market prices for meats and 'native' produce, restaurant reports, and fees levied against the storage of imported foods, markets, and slaughterhouses, as well as smuggling of foods, and strike actions by food workers in the early twentieth century. The ARG series was useful in learning about issues and conflicts related to market policing, town layouts, and general food related administration from the years 1900–1965.

The Records Assistant in Kumasi also suggested I review the ERG series, a methodical examination of which proved helpful for food related information after independence. The National Food and Nutrition Board and its recommendations for nutritionally balanced consumption feature prominently in the ERG series. A review of the ERG series also yielded much information about attempts by the Nkrumah government to initiate a packaged food industry with local fruits, livestock and milk production, and the attempted mechanization of millet, and rice production. Marketing and distribution systems reports were found in this series. I carefully reviewed these records, comparing them to my fieldwork research to build up an understanding of how the food distribution systems have changed over time in Ghana.

Along with archival records, I relied upon published primary sources and ethnographies to enable an interdisciplinary, historiographical approach throughout the work as the best means of accommodating a range of relevant sources at different levels of detail over a two-century period. For the first chapter it was necessary to construct an understanding of the geography, the staple starches, and the people who migrated to the Gold Coast. For this endeavour, both seventeenth- and eighteenth-century sources were consulted to build up a baseline of the eco-culinary zones—defined here as the microecologies and available foods. Especially useful in this undertaking for the coastal areas were the Danish and Dutch sources. The translation of the writings of Ludwig Ferdinand Rømer (1760), Johannes Rask (1754), and Hans Christian Monrad (1822) translated by Selena Winsnes, the translation of Pieter De Marées (1602) by Albert van Danzig and Adam Jones, and the translation of Willem Bosman (1705) were invaluable for building up a picture of the foodways there, especially in the coastal areas.[41]

Eurafrican Carl Christian Reindorf's *History of the Gold Coast and Asante* (1895) proved an especially valuable resource for the coastal areas and the forested centre in the nineteenth century.[42] His work furnished a local retelling of history and of how foods were used in a political and ritual context. Many of the rest of the sources for the nineteenth century coastal and forested regions were produced by British traders and missionaries, with the notable exception of Dutch African envoy, Willem Huydecoper (1817).[43] The work of anthropologists and district commissioners such as R. S. Rattray, Louis Tauxier, and A. W. Cardinall provided ecological and food information about both the central forested region and the Northern Region.[44]

---

[41] Ludewig Ferdinand Rømer, *A Reliable Account of the Coast of Guinea (1760)*, trans. Selena Axelrod Winsnes (Brooklyn: Diasporic Africa Press, 2013); Johannes Rask, *Two Views from Christiansborg Castle*, trans. Selena Axelrod Winsnes (Legon: African Books Collective, 2009); Hans Christian Monrad, *Two Views from Christiansborg Castle*, trans. Selena Axelrod Winsnes (Legon, Accra: African Books Collective, 2009); Pieter De Marées, *Description and Historical Account of the Gold Kingdom of Guinea (1602)*, trans. Albert van Dantzig and Adam Jones (Oxford: British Academy, 1987); William Bosman, *A New and Accurate Description of the Coast of Guinea* (London: Ballentyne, 1907), [London: James Knapton, orig. 1705].

[42] Carl Christian Reindorf, *History of the Gold Coast and Asante* (Basel: Missionsbuchhandlung, 1895).

[43] Willem Huydecoper, *Huydecoper's Diary: Journey from Elmina to Kumasi 28th April 1816–18th May 1817*, trans. Graham Irvine (Legon: University of Ghana Department of History, 1962).

[44] R. S. Rattray, *Ashanti Proverbs: The Primitive Ethics of a Savage People* (Oxford: Clarendon Press, 1916); R.S. Rattray, *Ashanti* (Oxford: Clarendon Press, 1923); R.S. Rattray, *Religion & Art in Ashanti* (Oxford: Clarendon Press, 1927); R. S. Rattray, *Ashanti Law and Constitution* (Oxford: Clarendon Press, 1929); Rattray, *Tribes*; Louis Tauxier, *Le noir de Bondoukou* (Paris: E. Leroux, 1921); Cardinall, *Natives of the Northern Territories*; A. W. Cardinall, *In Ashanti and Beyond* (London: Seeley, Service and Company Ltd., 1927); Cardinall, *The Gold Coast, 1931*.

Written by European traders, envoys, and missionaries, primary source discussions of cooking and eating were often an attempt to describe for educated European audiences who might travel to the area what the material culture was like. On the coast, the authors of this material focused on providing a general overview of what to expect from the locals there. Therefore, the particulars of the cultural aspects of foodways are often not addressed in detail. The same limitations apply to the work of the early twentieth-century ethnographers and administrators of the Northern Region.

The second chapter of the work is concerned with building up a picture of the cultural dictates of what would be considered a proper meal. The symbolism behind certain staple foods as exhibited in commensality habits (practices of eating or drinking together in a common physical setting), food preparation, and myths and legends are decoded. For this portion of the work, the sources were varied and interdisciplinary, making use of linguistic and anthropological research about food to understand how food has structured social organisation and reflected cosmological meanings. The work of social anthropologist Mary Douglas, structural anthropologist Claude Lévi-Strauss, and that of sociologist Pierre Bourdieu were especially helpful in the construction of a general framework of Gold Coast human culture and symbolism.[45] The research of anthropologists M. J. Field and Gracia Clarke, as well as that of sociologist Fran Osseo-Asare, were invaluable in analysing the meanings applied to food in the Ghanaian context. Reindorf and Rattray were indispensable sources for the myths and legends examined for links between food and its relationship to identity.

[45] Mary Douglas, *Purity and Danger: An Analysis of Concepts of Pollution and Taboo* (London: Routledge, 1966); Mary Douglas, *Natural Symbols: Explorations in Cosmology* (London: Barrie & Jenkins, 1973); Claude Lévi-Strauss, *The Raw and The Cooked: Introduction Science of Mythology* (Harper & Row, 1969); Pierre Bourdieu, *Language and Symbolic Power* (Cambridge: Polity Press, 1992); Pierre Bourdieu, *Distinction: A Social Critique of the Judgement of Taste* (Cambridge: Harvard University Press, 1984).

Fieldwork

This project makes use of fieldwork conducted in 2014 which consisted of interviews with government food researchers, tourism officials, port administrators, restaurant and store managers and chefs, catering school directors, and market and street food workers.[46] (See Appendix for the Questionnaire.) These informants provided information about attitudes toward what was considered a proper meal, issues concerning the use of imported foods, attitudes toward the concept of a national dish, and technical information about how food was prepared and who did the preparation. I met and interviewed people in the capital, Accra, as well as in Takoradi, Aburi, Tamale, and Bolgatanga. Research was conducted using both interviews of individuals and participant observation of public spaces such as government agencies, restaurants, and street food vendors.

Research was conducted using both interviews of individuals and participant observation of in places such as government agencies, restaurants, and street food vendor stalls. The individuals interviewed exemplified a broad cross section, 55% being male and 45% female, with uniform representation across various age groups. I conducted 62 interviews in Accra, Bolgatanga, Tema, Cape Coast, and points in between. Many interviewees, especially in Accra, were of mixed ethnic backgrounds with 9% self-identifying as of Asante ancestry, 23% of the Volta region, 13% from both the Greater Accra and Easter Region, and 17% foreigners. People interviewed represented a variety of Ghanaian ethnic groups, with nearly equal numbers of individuals from the Volta Region, the Greater Accra Region, the Eastern Region, and foreigners from Lebanon, the Philippines, and other West African countries. In terms of occupation, individuals were involved in business concerns, and employed in educational institutions. Clergy members, community leaders, and political leaders were interviewed. These persons, in addition to those employed in the arts or in the museum field, in positions in such diverse fields as public health, philanthropy, and urban studies, were all classified

[46] Brandi Simpson Miller, 'Food And Nationalism In An Independent Ghana' (MA thesis, Georgia State University, 2015), 90–97.

as professionals. All other interviewees were either classified as students or blue-collar workers.

Field work was limited to observing major markets, cold storage businesses, grocery stores, roadside food sales, ports, restaurants, hotels, resorts, and included various interviews with workers at government ministries and Non-Governmental Organizations. On only one occasion was I privileged with an opportunity to share a meal in a private home. No food events associated with any religious occasions were observed. Additionally, there was not sufficient time to make a first-hand observation of any food taboos or prohibitions.

My initial hypothesis was that there was a national cuisine currently under construction which had been heavily influenced by Ghana's historical relationship with the African American intelligentsia. This hypothesis was proven false, and I discovered that the opinions over the meaning of a national cuisine were as rich and wide-ranging as the varieties of regional foods themselves. Along with discussions of the regional differences in preparation techniques and ingredients, were heartfelt expressions of the roles food played in private life, in the Ghanaian economy, and in the messages their foods convey to visitors and citizens of the nation. Does a country need a national cuisine to be a great nation? What is the national cuisine of Ghana? When taken together the answers to these and other questions create a fascinating picture of how, when viewed through the lens of cuisine, Ghanaian cultural traditions have remained regional since independence. General trends found in the responses to the survey questions included expressions of pride in the wholesomeness of the food and its ability to satisfy hunger. Of frequent concern was the lack of consistent premarket oversight, and the scant monetary support for the local food industry. An appreciation of the diversity of the regional cuisines was also evident in many of the responses (Fig. 1.2).

The fieldwork revealed that Ghanaian food was served and preferred by most of the population, with continental cuisine reserved for formal state functions in the state-owned hotels. Western food such as frozen yogurt from Danish exporters or Maggi seasoning cubes were widely available for purchase on the street and in stores and markets. Most western foods, such as pizza or sandwiches, were widely considered to be snack food and not substantial enough to be served as a meal. Popular dishes from other

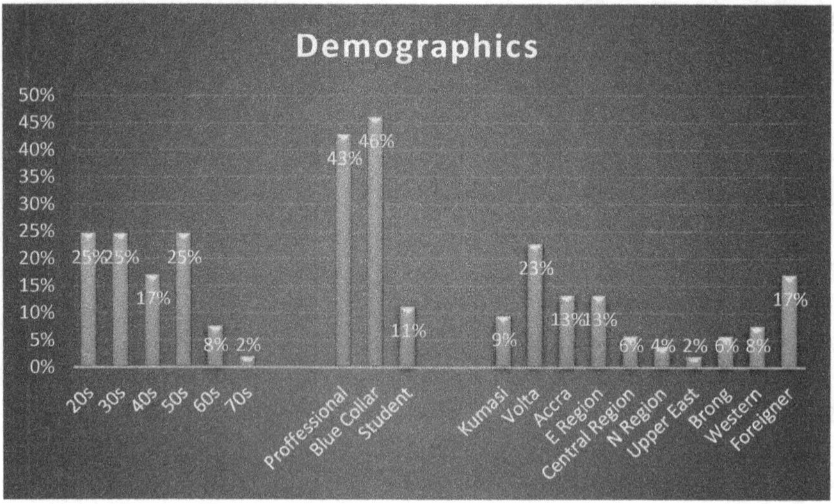

Fig. 1.2 Field work demographics

ethnic groups available on the road in rest stops and in urban centres included *fufu, banku, or tuo zafi* (different varieties of starchy dumplings), with fufu being the most popular. A dense, starchy staple made of pounded yam, fufu is primarily known as an ethnically Asante food.

Regional foodways in Ghana are remarkably similar in that customarily a substantial meal is served each day that consists of a large starchy dumpling, accompanied by a soup that is flavoured with various meats. There was traditionally little distinction made between course and mealtimes in the countryside. In urban centres people may have tea and toast or biscuits (cookies) for the morning repast reserving the main, and at times only meal, for later in the day. Fish was widely consumed, particularly in the south near the coast. No porkwas observed being served and chicken was considered a delicacy one encountered most often in urban centres. There was little beef available in the south, as most all beef was raised in the north. Goats were found foraging all over the countryside and in the cities like Accra and Kumasi. In the areas surrounding Nima market in Accra, which was heavily populated with Muslim northerners, a number of cattle are kept adjacent to people's homes in small shelters.

Neighbourhoods in Accra also feature large community ovens and street food cooking operations.

There are currently no efforts to construct a national cuisine in Ghana. There was an awareness of the concept of a national cuisine, but it did not seem to be a pressing concern as most people were apprehensive over fluctuating costs of food due to the unstable economy. Cookbooks tended to veer toward the preservation of regional cuisines and techniques. Those that featured European cuisine were intended as instructional manuals in a course of culinary study at technical colleges or training programs and are very different from regional cookbooks. The state tendency to spatially situate Ghanaian regional cuisines in deference to continental cuisine does not in fact indicate any perceived lack of pride in local cuisines. On the contrary, this special differentiation reflects a desire to make foreign guests feel comfortable (Figs. 1.3).

Not only was availability a priority, but the encouragement of entrepreneurialism and the development of new products for local foods was a priority as well. The following story is illustrative of the importance of

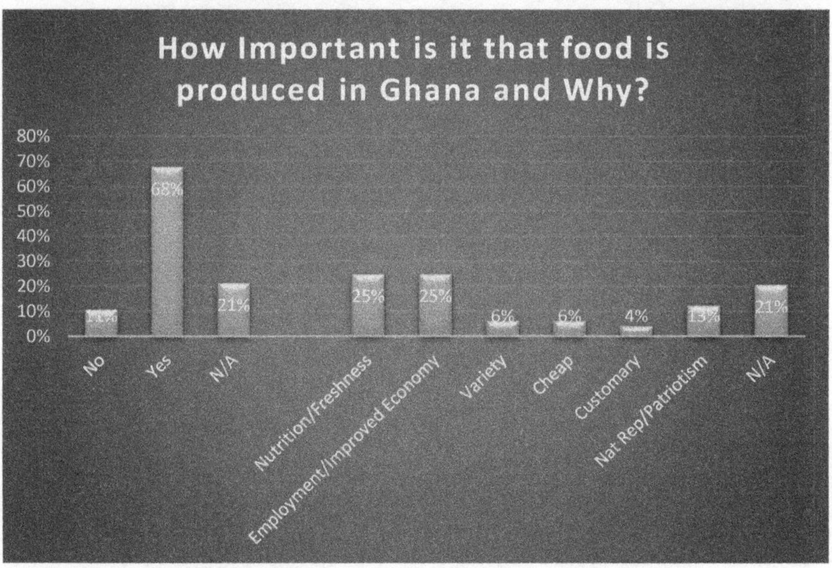

Fig. 1.3 Sample survey question

local foods and their development in Ghana. Halfway through my stay in Accra, I got the opportunity to interview Richard Twumbaah at the compound which formerly housed the UN food testing institute. Since then, the area had been taken possession of by the Ghanaian government and turned into the Food Research Institute (FRI). The current FRI office had since been moved to another location. The compound was still inhabited by many of the employees of FRI, however. Mr. Twumbaah, the security man who had been employed at this very same location since he finished school in 1968, shared with us the history of this location.

Twumbaah, an Akan from the Eastern Region, was one respondent who claimed the national dish of Ghana was rice since "At any party that is what is served."[47] Not only was he of the opinion that the government did not promote the food of Ghana but that the government "…can't do it as each tribe has its own food." He relayed to us that most anything he desired to purchase was available: "Yes, if you have money. At times it is too costly. You cannot afford your favourite food each day." Fufu was his favourite food which he gets an opportunity to eat once a week. He was the first person to tell us that there was such a product as fufu powder, which has been developed by FRI. He added that the government could do more to ensure the security of the food as it is brought directly from the village without inspection or oversight. After the interview, Mr. Twumbaah was kind enough to show us to the home of the Director of the Food Research Institute (Fig. 1.4).

At the Food Research Institute, I had the opportunity to interview three top officials who relayed their organizational mission to me, who distributed a copy of the annual report, and provided me with a sample of one of their products (palm nut pulp). They answered my questions and gave me a tour of their facilities, showing me the bio toxin testing labs, new products they developed, and their test kitchens. I was treated to a sample of bread made with yam flour. This was followed by a visit to the marketing department where there were many products they developed on display, such as powdered fufu. One of the mandates of FRI is to work with entrepreneurs to find different ways to develop, package, and market local foods. Interviews at FRI confirmed evidence of the

[47] Richard Twumbaah. Interview with author. Personal interview. Accra, June 2, 2014.

Fig. 1.4 FRI convenience products made of local starches

reinforcement of local production, a mandate to develop new products, and disinterest in the vernacularisation of cuisine in Ghana.

Dr. Atikpo, Deputy Director of FRI, was asked about the African American influence on Ghanaian foods. She dismissed this query with the following: "Our national foods are already established and known all over West Africa. They are delicious, popular, and varied, more than others in the region." When asked how much foreign food has displaced local food, Dr. Atikpo revealed that there currently were no foreign food policies as foreign foods 'are not feared by the Ghanaian government'. What her organization's responsibility was to oversee food's wholesomeness and quality. She estimated that roughly eighty-five percent of Ghanaians ate local food.[48] Her organization was one of thirteen nationwide that researched foods to advise the Food Ministry on policy for such items as palm oil, timber, rice, and maize. Although she admitted to Ghanaians having developed a taste for imported American rice, luncheon meat, and

[48] Dr. Margaret Ottah Atikpo, Deputy Director of the National Food Research Institute Ghana. Interview by author. Personal interview. Accra, June, 2, 2014.

cooking oil from food aid during the 1983 famine, she insisted that her organization promoted locally produced food, such as brown rice through the Africa Rice Project for export to places like Brazil and China. The choices of foodstuffs for Ghanaians owes as much to the Ghanaian government's policies as to tastes developed from famine induced foreign food aid.

Theoretical Underpinnings

Using an anthropological lens, this work explores identities embedded in the cuisines of the people of Ghana. By incorporating aspects of anthropological theories, it hopes to develop a deeper understanding of the historical cultural context and a more nuanced view of the social history of cooking and eating in Ghana. I have depended heavily on structuralist theory throughout the work. As indicated, by examining the ritualised nature of everyday food preparation and consumption, I am drawing here upon the work of Claude Levi Strauss and Mary Douglas. This approach allows for an engagement with the history of the broader social, political, and economic contexts of food production, preparation, and consumption. The employment of structuralist theory facilitated an understanding of how traditions have been remade in the contexts of new foods, new markets, and new political realities in Ghana.

Some post-structuralist theoretical elements have also been included in this work. Dictionaries in the local languages and linguistic analysis were consulted to provide an understanding of language and its role in creating food meanings. The nineteenth-century dictionaries by Basel Mission scholars were particularly useful, as was the more recent work of Blench and La Fleur.[49] The exploration of power relationships made visible

[49] Johann Zimmermann, *A Grammatical Sketch of the Akra or Ga Language* (Stuttgart: Basel Mission, 1858); J. G. Christaller, C.W. Locher, J. Zimmermann, *A dictionary, English, Tshi (Asànté), Akra* (Basel: C. Schultze, 1874); '"Observers Are Worried": A Dictionary of Ghanaian English', Roger Blench, last modified 12 January 2006, http://www.rogerblench.info/Language/English/Ghana%20English%20dictionary.pdf; La Fleur, *Fusion Foodways*.

through food in the work of Marcel Mauss and Martin Jones were notably beneficial for discussing the political use of food by the Asante in the nineteenth century.[50] Theories of embodiment and agency by Falk and Abbots were also employed throughout this work in order to convey the ways in which eating practices served as integrating rituals for the community.[51] Concepts of bodily self-control and discipline are not explored in this project. Rather, the focus is on the use of food as a commodity to reflect on one's identity and one's place in the world.[52]

Anthropologist Richard Wilk mused that in a world of 'constant cultural contact, international media, and marketing, the process of change in diets seems to have accelerated, but the boundaries that separate cultures have not disappeared.'[53] Wilk stated that since independence in 1981 Belizeans became aware of the need to craft a national Belizean identity, and that this cultural performance should include a national food. He argued that Belizean cuisine came out of a reaction to the foreign influence of the US and the UK. With 73% of the population having travelled or having close family members abroad, the colonial elite could no longer control the flow of information, resulting in a heightened awareness of the necessary performance of local authenticity through food—a performance that would demonstrate their worldliness, modernity, and sophistication.

Ghanaians have had access to foreign cultural goods for hundreds of years as they played a vital role in the Atlantic World trade. Decidedly cosmopolitan, many coastal people intermarried with Europeans and

---

[50] Marcel Mauss, *The Gift: Forms and Functions of Exchange in Archaic Societies* (London: London: Cohen & West, 1966); Martin Jones, *Feast: Why Humans Share Food* (Oxford: Oxford University Press, 2008).

[51] Pasi Falk, *The Consuming Body* (Newbury Park, SAGE Publications, 1994), 20; Emma-Jayne Abbots, *The Agency of Eating: Mediation, Food and the Body* (London: Bloomsbury Publishing, 2017), 36–42.

[52] Falk, *The Consuming Body*, 21; Psyche Williams-Forson, 'Other Women Cooked for My Husband: Negotiating Gender, Food, and Identities in an African American/Ghanaian Household', *Feminist Studies* 36, no. 2 (2010): 435–61; Psyche Williams-Forson, '"I Haven't Eaten If I Don't Have My Soup and Fufu": Cultural Preservation through Food and Foodways among Ghanaian Migrants in the United States', *Africa Today* 61, no. 1 (2014): 68–87.

[53] Richard R. Wilk, '"Real Belizean Food": Building Local Identity in the Transnational Caribbean', *American Anthropologist* 101, no. 2 (1999): 276.

educated their children abroad.[54] Ghana became independent almost a quarter of a century before Belize and yet, their regional foods remain largely 'hidden' to outsiders. Unlike Belizeans, Ghanaians do not expect you to seek out local cuisine, do not 'perform' a national cuisine for tourists, and are surprised when tourists express any knowledge of or desire to eat local dishes. An historical cultural diglossia has coexisted with regional foodways based on a starchy staple before, during, and after colonization. This is evidenced by the development and growth of urban chop bars (restaurants that served regional dishes) in the nineteenth century (Chap. 4). Not until the last few years have Ghanaian local foods been revealed and shared with tourists in high-end local food restaurants like Tatale or the Northern Plate, both located in Accra.

What is driving the attention toward showcasing regional cuisines in Ghana is the challenge to regional foods brought about by the inherent lifestyle changes in rapidly advancing urbanization. Some scholars argue that urbanization has had the biggest impact on diets.[55] People are living further away from work and family, and this has necessitated some efficiency with regard to meal preparation. Scholar Fran Osseo-Asare commented that people in Ghana are looking for labour saving shortcuts like Maggi seasoning cubes versus free range meats and beans. She stated that the portability and convenience of wheat has pushed its consumption up significantly since 2005.[56] Indeed, from 2015 to 2019 imports of easy to cook rice have increased by 85%.[57]

The consumption of convenience foods in places like Belize has also grown significantly over the last few decades: 'As Belizeans themselves become more cosmopolitan diners, they also start to demand more

[54] Wulf Joseph Wulf, *A Danish Jew in West Africa: Wulf Joseph Wulff Biography And Letters 1836–1842*, trans. Selena Axelrod Winsnes (Sub-Saharan Publishers, 2014); Monrad, *Two Views*; Reindorf, *History*; Huydecoper, *Diary*; Thomas Birch Freeman, *Journal of Various Visits to the Kingdoms of Ashanti, Aku, and Dahomi, in Western Africa* (London: John Mason, 1844), 161.

[55] Jessica Ham (Assistant Professor of Anthropology, Oxford College, Emory University), interviewed by Brandi Simpson Miller, via telephone, April 21, 2021, Abingdon, Oxfordshire, UK.

[56] Barbara Baeta (Proprietor, Flair Catering, Cantonments, Accra, Ghana), 'Brandi Miller's requested interview,' interviewed by Fran Osseo-Asare, Flair Catering, Cantonments, Accra, March 22, 2018.

[57] 'Country Briefs—Ghana,' GIEWS—Global Information and Early Warning System, FAO, accessed May 2021, http://www.fao.org/giews/countrybrief/country.jsp?code=GHA.

variety that can ever be produced at home, including all the complex diet foods and prepared frozen convenience dishes that the U.S. food industry produces in such profusion.'[58] Similar lifestyle changes in Ghana, brought about by urbanization, resemble a sort of creolization—a process Wilks characterizes as a 'global mixture' of local and imported foods. Packaged foods, imported grains, and precooked local staple starches are being incorporated into the local repertoire.[59] The submersion of bouillon stock cubes into the traditional soup as a shortcut to boiling your own soup stock is the most common expression of this global creolization in Ghana. Wilks sees the 'compression' of different western cuisines into one as part of this 'global mixture' process. In Ghana compression can be seen in the serving of 'continental' food in upscale restaurants, hotels, and at state banquets, continental food being a blending of different western dishes. This sort of global creolization is becoming increasingly common in Ghana. Though I concede that many of the same processes Wilk elaborates upon in his monograph about Belize are also happening in Ghana, the result is still not the performance of a national cuisine. Instead, regional dishes remain a fixture of the culinary landscape with no typical national cuisine in the process of development.

Despite the spread of global creolization to Ghana as seen in the consumption and submersion of foreign starches and packaged foods, the sense is that people in Ghana *want to be eating traditional food.*[60] Ghanaians value their regional cuisines for the work they do in representing and replicating their ethnic identity. What is more, according to this view the consumption and promotion of regional foods does triple duty—regional cuisines are closely tied to cultural identity, with the preservation of the environment, and the protection and bolstering of the national economy. Take for example The Ghana Food Movement (GFM). GFM started in response to the perceived lack of regional cuisines and ingredients found in urban centres. Founded upon the belief that promoting local food production is good for the economy, the environment and for the preser-

[58] Richard Wilk, *Home Cooking in the Global Village: Caribbean Food from Buccaneers to Ecotourists* (Oxford: Berg, 2006), 186.

[59] Wilks, *Home Cooking*, 109.

[60] Jessica Ham, interview.

vation of local identities, GFM seeks to generate dialogue about the lack of diversity in urban diets by promoting offline meetups to eat delicious local food.[61] They also employ the use of social media platforms like Facebook and YouTube to promote regional cuisine and to showcase regional food eateries that would appeal to tourists.[62]

The perception of a growing threat to regional foodways in urban centres is echoed in the sentiments of people who commented on the recent opening of the Domino's Pizza branches. On 22 March of 2021, one respondent wrote 'The Ghanaian FDA Must Wake Up':

Does Ghana have a proper restaurant inspection system? Just after Coming To America movie 2, they all wanna start launching in Africa with all these unhealthy foods. Ghana's Asanka Locals and the rest must be supported heavily. No tourist will visit Ghana because of Dominos Pizza restaurant, but rather a Kontomere with Yam or Banku and Tilapia fast food restaurant. Branding and execution is what Ghana or perhaps Africa needs. It's the twenty-first century, the technology and the technical know how is right here. Arise Ghana youth for your country![63]

This patriotic comment demonstrates that not only do some believe that fast foods imperil regional identity and traditional food knowledge, but they also jeopardize Ghana's existence 'as a politically independent people.'[64]

This remark and others like it indicate there is considerable tension over the 'launching in Africa' of 'unhealthy foods' which is then connected to the undermining of political autonomy. How does the introduction of a pizza chain threaten national sovereignty? The regional starchy staple in Ghana play a role similar to that of rice in Japan—as

---

[61] 'FAQ,' Ghana Food Movement, accessed May 2021, https://www.ghanafoodmovement.com/faq.html.

[62] Davidsbeenhere, 'Spicy GHANA FOOD TOUR!! Tatale, Egusi & Abolo,' YouTube, accessed May 2021, https://youtu.be/3D_tgwzwKO0; Davidsbeenhere, 'Northern GHANA FOOD in Accra!! Tuo Zaafi, Wagashi & Kuli Kuli,' YouTube, accessed May 2021, https://youtu.be/RvY2XwjoX3c.

[63] 'Domino's Pizza, world's largest pizza company, set to launch in Accra,' Ghana Web, last accessed May 11, 2021, https://www.ghanaweb.com/GhanaHomePage/NewsArchive/Domino-s-Pizza-world-s-largest-pizza-company-set-to-launch-in-Accra-1211608?gallery=1.

[64] 'About Us,' Food Sovereignty Ghana, accessed May 2021, https://foodsovereigntyghana.org/about-us/.

representing the collective self and marking the cultural boundary between that self and others.[65] The staple food is laden with symbolism, embodying regional space in the land where the starch is grown, and time in the rhythm of the growing and harvesting cycle. Planting and harvesting rhythms come to embody a people's history. The staple is a dominant metaphor for the social group and the way they define and redefine themselves from other Ghanaians and westerners (and lately, easterners). The symbolism of regional starchy staples was bifurcated; the starch was 'our food' and the earth it grew out of was 'our land,' shaping the regional social order in the form of cosmology, commemorative festivals, and foods.

The rapid growth of fast-food chains—at an annual rate of 20%—has forced Ghanaians to examine their consumption of fast food, which has been linked to health problems like obesity.[66] Encounters with other cultures prompt people to think about who they are in relation to others. "Tell me what you eat and I will tell you what you are" said Jean Anthelme Brillat-Savarin, the famous French epicure and gastronome.[67] Regional starchy staples are a way to think of the self in relation to others. Since at least the early nineteenth century we have heard Ghanaians, themselves valorise agricultural production (Chap. 4). Advocacy for traditional regional foods and ways of eating in Ghana have broadened in the last 20–30 years due to urbanization and growing access to packaged and fast foods. With advocacy groups like GFN, the production of traditional regional starches is being connected to the preservation of the environment and as a solution to climate change. Japan has undergone much the same sort of advocacy where proponents asserted that rice production preserved the soil, the air and other vegetation. In these discussions, the starch itself symbolizes a specific ethnicity as well as that groups' land, water, and air. The land itself is a metaphor for a pristine self, and the growing introduction of fast food, genetically modified seeds, and

[65] Emiko Ohnuki-Tierney, *Rice as Self: Japanese Identities through Time* (Princeton: Princeton University Press, 1993), 3–10, 109.

[66] James Boafo, 'Ghanaians are eating more fast food: the who and the why,' The Conversation, accessed May 2021, https://theconversation.com/ghanaians-are-eating-more-fast-food-the-who-and-the-why-153810.

[67] Jean Anthelme Brillat-Savarin, *The Physiology of Taste: Or Meditations on Transcendental Gastronomy: Transcendental Gastronomy* (Scotts Valley: CreateSpace Independent Publishing Platform, 2017), 25.

fertilizer inputs is considered impure and a threat to the regional self and to the nation's sovereign ability to produce its regional starchy staples.[68] The nation must preserve the ability to produce, process, cook and eat the regional starchy staple in its everyday home cooking.

Jack Goody observed that little distinction was made in Ghana between the food of cultural elites and home cooking.[69] Wilk reflected on the work of Jack Goody in his making of distinctions between polarities in food culture in cooking versus cuisine, cooking being the quotidian lived practice versus the elite or *haute* performative aspect of cuisine.[70] As food was/is so closely identified with ethnicity and region in Ghana, performative cuisine there would resemble serving *more* of your own quotidian ethnic food, not less: 'Rice is the defining feature of 'traditional Japanese cuisine'...What makes any dish *washoku* [Japanese cuisine], however, whether *haute* or not, is the presence of rice, no matter how small.'[71] Everyone in Japan ate the starchy staple regardless of class, and the same was true in Ghana. This was because rulers' power *came from their proximity to and service of* their dependents. Status in West Africa typically was/is contingent upon how many dependents one could feed. Therefore, when it came to feasting, the performance was *more regional starch* instead of a distinct hybrid elite cuisine. Public performance then was about how many dependents you could serve with the food that made you a Fante, Gã, or Asante. Therefore, serving foreign food to signal an elite, *haute* cuisine would be self-defeating. Evidence of this can be seen by the discussion in Chap. 3 where the Asantehene (Asante King) took great care to ensure that Europeans were served foods familiar to them when they feasted at the Asante court. The Asantehene was making evident his power to feed these foreigners, dependent upon him for their sustenance, with the food that materially recreated their identities.

Since the fifteenth century, the starchy staple of different ethnic groups in Ghana has materially changed due to lengthy experimentation, migration, or conflict. During these historical processes, the regional starchy

---

[68] Ohnuki-Tierney, *Rice as Self*, 105–11.

[69] Jack Goody, *Cooking, Cuisine and Class: A Study in Comparative Sociology* (Cambridge: Cambridge University Press, 1982).

[70] Wilk, *Home Cooking*, 106.

[71] Ohnuki-Tierney, *Rice as Self*, 108.

staple continued to represent the regional self/ethnic group while it underwent various historic changes. As Tierney stated: 'This double linkage to the Japanese self—representing its *body* and its *land*—may be a clue to the enormous power and resilience of rice symbolism *which has remained more meaningful and powerful than both the imperial system and agriculture itself*.'[72] Similarly, in Ghana, linkages to *body* and *land* through the regional starchy staple are why Ghanaian cuisine has outlasted hundreds of years of trade, colonization, and will doubtlessly endure through the challenges of urbanization and climate change. Even though consumption of imported starches has risen steadily since the nineteenth century, the symbolism of the regional starchy staple as a metonym for the body, and its links to space and time through the land and its history have proved both enormously powerful and resilient.

This comparison of Ghanaian foodways to historical developments in both Belize and Japan is illustrative of the deep symbolic relevance of the regional starchy staple which is the foundation of Ghanaian foodways. This projects exploration of the evolution of this symbolism contributes to the study of culinary history as a thorough examination of the multiple factors that affect nation-building in West Africa, and how and why people and governments make culinary choices. In a series of case studies to examine key moments of food innovation and food consolidation this research will show how certain foodways became "traditional" main dishes across three regions of Ghana. As yet, no work has attempted this sort of a detailed approach to the study of Ghana, and as such this work will contribute to comparative studies with other countries and times, both in Africa and beyond.

Organisation of Chapters

The first two chapters use anthropological material as well as oral histories to establish a framework for the particularities of geographies, languages, historical migrations, and ethnic groups. Chapters 3, 4, and 5 are organised chronologically and by each of the three major eco-culinary

[72] Ohnuki-Tierney, 132.

zones—the forested centre, the coastal region, and the savanna north. Using first-person accounts, government documents, proverbs, and ethnographies, these chapters explore the foodways of each eco-culinary zone and consider how people invested food with meaning through seasonal rhythms and ritual.

Chapter 3 explores themes of power, differentiation, and separation in the great forest kingdom of Asante. Utilising a comparative approach, case studies of the accounts of European envoys to the Asante capital of Kumasi are analysed. The differences in the food gifts received by the envoys from the monarch are compared and contextualised in the political realities of the time. Variations in commensality customs and feasting over time are also explored in this chapter.

Chapter 4 establishes a baseline for the foodways of the coastal Fante, Gã, and Ewe in the nineteenth century, and examines how that baseline was disrupted as a result of the transition away from the slave trade. Consulting published primary source accounts of traders, missionaries, and administrators, this chapter elucidates the differences in foodways between the three groups. Ecological factors and the push factors of expanding internal empires are traced through what was cooked and eaten. The beginnings of the incorporation of imported foods are studied as well.

Chapter 5 looks at the late nineteenth and twentieth-century foodways of the savanna region, focusing on the seasonality of cooking and eating there. The use of ritual and commensality in building community is also a theme explored in this chapter. Ethnographies conducted by both district commissioners and scholars, as well as government reports, are used to compose a region wide normative cuisine.

Chapter 6 bridges together the changes over time from the nineteenth to the twenty-first century for the whole of Ghana. Fieldwork, newspapers, government documents, and crop studies bring the analysis up to the current day. Attitudes toward consumption and how the modern nation of Ghana cooks and eats are explored.

The six chapters of this work cannot claim to be a comprehensive history of what was cooked and eaten in the nineteenth- and twentieth-century Gold Coast. However, it is hoped that by bringing together a range of interdisciplinary sources, and by examining the three major eco-culinary zones and their differences and similarities, these chapters can contribute to a history of this kind. With this project, my hope is to impact the food studies field by making Ghana's culinary heritage visible in how it operates both internally, and in interaction via trade and exchange with the rest of the world. Food studies is an amazingly varied discipline with a plethora of approaches. By providing a baseline from which future scholars can begin their investigations, this project contributes toward backfilling the hole in the food studies scholarship of Africa.

References

Primary Sources

M.A. Atikpo Dr., Deputy Director of the National Food Research Institute. Interview with author. Accra, 2 June 2014

B. Baeta, Founder and Owner of Flair Catering Services. Interviewed by Fran Osseo-Asare. Flair Catering, Cantonments, Accra, Ghana, 22 March 2018

W. Bosman, *A New and Accurate Description of the Coast of Guinea* (Cass & Co., London, 1967)

A.W. Cardinall, *The Natives of the Northern Territories of the Gold Coast; Their Customs, Religion and Folklore* (Routledge, London, 1920)

A.W. Cardinall, *In Ashanti and Beyond* (Seeley, Service and Company Ltd., London, 1927)

A.W. Cardinall, *In Ashanti. The Gold Coast, 1931* (Government Printer, Accra, 1932)

Country briefs—Ghana, GIEWS—Global Information and Early Warning System, FAO, http://www.fao.org/giews/countrybrief/country.jsp?code=GHA. Accessed May 2021

P. De Marees, *Description and Historical Account of the Gold Kingdom of Guinea (1602)*. Translated and edited by A.v. Dantzig, A. Jones (British Academy, Oxford, 1987)

M. J. Field, Gold coast food. *Petits Propos Culinaire*, no. 43, 7–19 (1993) [orig. *Gold Coast Food* (Achimoto College Press, Achimoto, 1931)]

T.B. Freeman, *Journal of Various Visits to the Kingdoms of Ashanti, Aku, and Dahomi, in Western Africa* (J. Mason, London, 1844)

J. Ham, (Assistant Professor of Anthropology, Oxford College, Emory University.) Interview by author via telephone. Abingdon, Oxfordshire, UK, 21 April 2021

W. Huydecoper, *Huydecoper's Diary: Journey from Elmina to Kumasi 28th April 1816–18th May 1817*. Translated by W. Graham (University of Ghana, Legon, 1962)

H.C. Monrad, *Two Views from Christiansborg Castle*. Translated by S.A. Winsnes (African Books Collective, Legon, 2009)

J. Rask, *Two Views from Christiansborg Castle*. Translated by S.A. Winsnes. (African Books Collective, Legon, 2009)

R.S. Rattray, *Ashanti Proverbs: The Primitive Ethics of a Savage People* (Clarendon Press, Oxford, 1916)

R.S. Rattray, *Ashanti* (Clarendon Press, Oxford, 1923)

R.S. Rattray, *Religion & Art in Ashanti* (Clarendon Press, Oxford, 1927)

R.S. Rattray, *Ashanti Law and Constitution* (Clarendon Press, Oxford, 1929)

R.S. Rattray, *The Tribes of the Ashanti Hinterland* (Clarendon Press, Oxford, 1932)

C.C. Reindorf, *History of the Gold Coast and Asante* (Missionsbuchhandlung, Basel, 1895)

L.F. Rømer. *A Reliable Account of the Coast of Guinea (1760)*. Translated by S.A. Winsnes (British Academy, London, 2000)

L. Tauxier, *Le noir de Bondoukou* (E. Leroux, Paris, 1921)

R. Twumbah, Interview with author. Accra (2 June 2014)

J.W. Wulf, *A Danish Jew in West Africa: Wulf Joseph Wulf Biography and Letters 1836–1842*. Translated by S.A. Winsnes (Sub-Saharan Publishers, Accra, 2014)

Secondary Sources

E.-J. Abbots, *The Agency of Eating: Mediation, Food and the Body* (Bloomsbury Publishing, London, 2017)

E. Akyeampong, *Drink, Power, and Cultural Change: A Social History of Alcohol in Ghana, c. 1800 to recent times* (Heinemann, Portsmouth, 1996)

M.L. Arnott, *Gastronomy: The Anthropology of Food and Food Habits* (Walter De Gruyter Mouton, Berlin, 2011)

G. Austin, J. Baten, B. Van Leeuwen, The biological standard of living in early nineteenth-century West Africa. Econ. His. Rev. **65**(4), 1280–1302 (2012)

BBC, The misadventures of Romesh Ranganathan, Series 1, Episode 2, https://www.bbc.co.uk/programmes/b0bb3gnc. Accessed 8 June 2020

D. Bell, G. Valentine, *Consuming Geographies: We Are Where We Eat* (Routledge, London, 1997)
BETUMI, African cuisine, http://www.betumi.com/home/gumbo-fulltext.html. Accessed 31 July 2018
BetumiBlog: Ghana-style Kenkey, https://www.betumi.com/2007/03/ghana-style-kenkey-italy-has-polenta.html. Accessed 18 May 2020
R. Blench, *Natural Resource Management in Ghana and Its Socio-Economic Context* (Overseas Development Institute, London, 1999)
R. Blench, M. Dendo, "Observers are worried": A dictionary of Ghanaian English. Ghana English dictionary (2006), http://www.rogerblench.info/Language/English/Ghana%20English%20dictionary.pdf
J. Boafo, Ghanaians are eating more fast food: the who and the why. The Conversation, https://theconversation.com/ghanaians-are-eating-more-fast-food-the-who-and-the-why-153810. Accessed May 2021
P. Bourdieu, *Distinction: A Social Critique of the Judgement of Taste* (Harvard University Press, Cambridge, 1984)
P. Bourdieu, *Language and Symbolic Power* (Polity Press, Cambridge, 1992)
J.A. Brillat-Savarin, *The Physiology of Taste: Or Meditations on Transcendental Gastronomy: Transcendental Gastronomy* (CreateSpace Independent Publishing Platform, Scotts Valley, 2017)
C. Camp, *American Foodways: What, When, Why and How We Eat in America* (August House Publishing Inc, Little Rock, 1989)
J.A. Carney, *Black Rice: The African Origins of Rice Cultivation in the Americas* (Harvard University Press, Cambridge., 2002)
J. Carney, R. Rosomoff, *In the Shadow of Slavery Africa's Botanical Legacy in the Atlantic World* (University of California Press, Berkeley, 2009)
J.-P. Chretien, The historical dimension of alimentary practices in Africa. Diogenes **36**(144), 92–115 (December 1988). https://doi.org/10.1177/039219218803614406
J.G. Christaller, C.W. Locher, *A dictionary, English, Tshi (Asànté), Akra* (Evangelical Missionary Society, Basel, 1874)
G. Clark, *Onions Are My Husband: Survival and Accumulation by West African Market Women* (University of Chicago Press, Chicago, 1994)
G. Clark, From fasting to fast food in Kumasi, Ghana, in *Food Consumption in Global Perspective: Essays in the Anthropology of Food in Honour of Jack Goody*, ed. by J. Klein, A. Murcott, (Palgrave Macmillan, Basingstoke, 2014)
A. Crosby, *The Columbian Exchange: Biological and Cultural Consequences of 1492* (Greenwood Publishing Company, Westport, 1972)

I. Cusack, African cuisines: recipes for nation-building? J. Afr. Cul. Stud. **13**(2), 207–225 (December 2000)

Davidsbeenhere, Spicy GHANA FOOD TOUR!! Tatale, Egusi & Abolo. YouTube, https://youtu.be/3D_tgwzwKO0. Accessed May 2021

Davidsbeenhere, Northern GHANA FOOD in Accra!! Tuo Zaafi, Wagashi & Kuli Kuli. YouTube, https://youtu.be/RvY2XwjoX3c. Accessed May 2021

A. Davidson, *The Oxford Companion to Food. The Oxford Companion to Food* (Oxford University Press, 1999)

A. Dede, *Ghanaian Favourite Dishes* (Anowuo Educational Publications, Accra, 1969)

J. Destombes, Nutrition and chronic deprivation in the West African Savanna: North Eastern Ghana, c. 1930–2000. PhD diss., London School of Economics and Political Science, 2001

J. Destombes, From long-term patterns of seasonal hunger to changing experiences of everyday poverty: Northeastern Ghana c. 1930–2000. J. Afr. His. **47**(2), 181–205 (2006)

M. Douglas, *Purity and Danger: An Analysis of Concepts of Pollution and Taboo* (Routledge, London, 1966)

M. Douglas, *Natural Symbols: Explorations in Cosmology* (Penguin, Harmondsworth, 1973)

R.E. Dumett, *Survey of Research Materials in the National Archives of Ghana* (Gemund, Lempp Verlag, 1974)

P. Falk, *The Consuming Body* (SAGE, Newbury Park, 1994)

J.-L. Flandrin, M. Montanari (eds.), *Food: A Culinary History from Antiquity to the Present*. Translated by A. Sonnenfeld (Columbia University Press, New York, 1999)

Food Sovereignty Ghana, About us, https://foodsovereigntyghana.org/about-us/. Accessed May 2021

M. Fortes, S.L. Fortes, Food in the domestic economy of the Tallensi. Africa **9**(2), 237–276 (1936)

Ghana Food Movement, FAQ, https://www.ghanafoodmovement.com/faq.html. Accessed May 2021

Ghana Web, Domino's Pizza, world's largest pizza company, set to launch in Accra, https://www.ghanaweb.com/GhanaHomePage/NewsArchive/Domino-s-Pizza-world-s-largest-pizza-company-set-to-launch-in-Accra-1211608?gallery=1. Accessed 11 May 2021

J. Goody, *Cooking, Cuisine and Class: A Study in Comparative Sociology* (Cambridge University Press, Cambridge, 1982)

J. Goody, E.N. Goody, Food and identities: changing patterns of consumption in Ghana. Camb. Anthropol. **18**(3), 1–14 (1995)
J. Harris, *High on the Hog: A Culinary Journey from Africa to America* (Bloomsbury, New York, 2011)
D.P. Henige, The national archives of Ghana: a synopsis of holdings. Int. J. Afr. Hist. Stud. **6**(3), 475–486 (1973)
E.B. Ikpe, *Food and Society in Nigeria: A History of Food Customs, Food Economy and Cultural Change, 1900–1989* (Franz Steiner, Stuttgart, 1994)
D. Johnston et al., Review: time use as an explanation for the agri-nutrition disconnect: evidence from rural areas in low and middle-income countries. Food Policy **76**, 8–18 (2018)
M. Jones, *Feast: Why Humans Share Food* (Oxford University Press, Oxford, 2008)
K. F. Kiple, K. C. Ornelas (eds.), *The Cambridge World History of Food* (Cambridge University Press, Cambridge, 2000)
J.D. La Fleur, *Fusion Foodways of Africa's Gold Coast in the Atlantic Era* (Brill, Leiden, 2012)
C. Lévi-Strauss, *The Raw and The Cooked* (Harper & Row, San Francisco, 1969)
A. Logan, A history of food without history: food, trade, and environment in West-Central Ghana in the second millennium AD. PhD diss., University of Michigan, 2013
A. Logan, "Why can't people feed themselves?": Archaeology as alternative archive of food security in Banda, Ghana. Am. Anthropol. **118**(3), 508–524 (2016)
E. Mandala, Beyond the "Crisis" in African food studies. J. Hist. Soc. **3**(3–4), 281–301 (2003)
M. Mauss, *The Gift: Forms and Functions of Exchange in Archaic Societies* (Cohen & West, London, 1966)
J. McCann, *Green Land, Brown Land, Black Land: An Environmental History of Africa, 1800–1990* (Heinemann, Portsmouth, 1999)
J. McCann, *Maize and Grace: Africa's Encounter with a New World Crop, 1500–2000* (Harvard University Press, Cambridge, 2007)
J. McCann, *Stirring the Pot: A History of African Cuisine* (Ohio University Press, Athens, 2009)
J. McCann, Writing on the African pot: recipes and cooking as historical knowledge, in *Writing Food History: A Global Perspective*, ed. by K. W. Claflin, P. Scholliers, (Berg, New York, 2012)
J. Miller, *Food Studies: An Introduction to Research Methods* (Berg, Oxford, 2009)

S.W. Mintz, *Sweetness and Power: The Place of Sugar in Modern History* (Penguin Books, 1986)

S.W. Mintz, C.M. Du Bois, The anthropology of food and eating. Annu. Rev. Anthropol. **31**, 99–119 (2002)

P. Nugent, Do nations have stomachs? food, drink and imagined community in Africa. Afr. Spectr. **45**(3), 87–113 (March 2011)

E. Ohnuki-Tierney, *Rice as Self: Japanese Identities through Time* (Princeton University Press, Princeton, 1993)

F. Osseo-Asare, *A Good Soup Attracts Chairs: A First African Cookbook for American Kids* (Pelican Pub Co Inc, La, 1993)

F. Osseo-Asare, *Food Culture in Sub-Saharan Africa* (Greenwood Press, Westport, 2005)

F. Osseo-Asare, B. Baëta, *The Ghana Cookbook* (Hippocrene Books, New York, 2015)

J.M. Pilcher, *Que Vivan Los Tamales!: Food and the Making of Mexican Identity* (University of New Mexico Press, Albuquerque, 1998)

J.M. Pilcher, *The Oxford Handbook of Food History* (New York, Oxford University Press USA, 2012)

F.M. Purcell, Nutrition in the colonial empire. BMJ **2**(4100), 294–296 (1939)

J. Rich, *A Workman Is Worthy of His Meat: Food and Colonialism in the Gabon Estuary* (University of Nebraska Press, Lincoln, 2007)

A. Richards, *Hunger and Work in a Savage Tribe: A Functional Study of Nutrition among the Southern Bantu* (George Routledge & Sons, London, 1932)

A. Richards, *Land, Labour and Diet in Northern Rhodesia: An Economic Study of the Bemba Tribe* (Oxford University Press, Oxford, 1939)

P. Shaffer, Seasonal Hunger in the northern territories of the gold coast, 1900–40. Econ. Hist. Dev. Reg. **32**(3), 270–300 (September 2, 2017)

B. Simpson Miller, Food and nationalism in an independent Ghana. MA Thesis, Georgia State University, 2015

B. Simpson Miller, Food and nationalism in an independent Ghana, in *The Emergence of National Food*, ed. by A. Ichijo, V. Johannes, R. Ranta, (Bloomsbury Academic, London, 2019)

R. Tannahill, *Food in History* (Stein and Day, New York, 1973)

R. Wilk, "Real Belizean food": building local identity in the transnational Caribbean. Am. Anthropol. **101**(2), 244–255 (1999)

R. Wilk, *Home Cooking in the Global Village: Caribbean Food from Buccaneers to Ecotourists* (Berg, Oxford, 2006)

P. Williams-Forson, Other women cooked for my husband: negotiating gender, food, and identities in an African American/Ghanaian household. Fem. Stud. **36**(2), 435–461 (2010)

P. Williams-Forson, "I haven't eaten if i don't have my soup and Fufu": cultural preservation through food and foodways among Ghanaian migrants in the United States. Afr. Today **61**(1), 68–87 (2014)

J. Zimmermann, *A Grammatical Sketch of the Akra or Ga Language* (Basel Mission, Stuttgart, 1858)

2

Ghana's Eco-Culinary Zones

In his pioneering *History of the Gold Coast and Asante* (1895), Rev. C. C. Reindorf began by explaining why he connected the history of the Asante with that of the Gold Coast. The British settlements on the coast became a formal colony in 1874. Asante, however, remained a sovereign kingdom, and was not occupied by the British until 1896 or formally annexed until 1901. The Northern Territories became a protectorate shortly after, in 1902.[1]

> There must be a starting-point in writing a history of a nation. If the kingdom of Akra, which appears to have been the first established on the Gold Coast, could have continued and absorbed that of Fante, or been absorbed by the latter, I might have easily obtained the starting-point. But both kingdoms having failed and the kingdom of Asante having become the leading and ruling power, a Gold Coast history would not be complete without the history of Asante, as the histories of both countries are so interwoven.[2]

[1] Roger S. Gocking, *The History of Ghana* (Westport: Greenwood, 2005), xxviii, 31.

[2] Carl Christian Reindorf, *History of the Gold Coast and Asante* (Basel: Missionsbuchhandlung, 1895), v.

B. Simpson Miller, *Food and Identity in Nineteenth and Twentieth Century Ghana*,
Food and Identity in a Globalising World,
https://doi.org/10.1007/978-3-030-88403-1_2

In this passage Reindorf made it clear that the Fante, Gã, and Asante were considered different polities, irrespective of their relative proximity. This was due in large part to the ecological differences between them, which in turn was reflected in their variances in language and in staple foods.

Local food is both ecologically determined and culturally defined.[3] A cuisine is a body of culturally defined practices, representations, and classifications.[4] The practices of a cuisine encompass both the physical and psychological transformation of food to render it safe to consume. A cuisine makes food representative of what is good to eat in a way that orders the world and gives it meaning. For example, in most western cultures, insects are not fit for eating. Last, a cuisine classifies when, with whom, and with what to eat something. Most westerners would not eat spaghetti for breakfast, for example. Thus, local foodways are culturally determined. However, there are ecological parameters that also give structure to a cuisine.

For patterns of food consumption to become a cuisine, a long-term biological relationship and a sustainable harvest needs to be in place. In this first chapter I characterize the ecological contributors to the cultural construction of cuisine in Ghana. Culture, ecology, and place are difficult to independently identify. I use language to reflect cultural relatedness and differences and use descriptions of biomes to reflect the ecological relatedness and divergences between language groups. Also incorporated is a discussion of how the arrival of new foods impacted those biomes and the cultures that inhabited them. From the arrival of the Portuguese in the 1470s, the Gold Coast was drawn into the nascent Atlantic economy which, by the sixteenth century, resulted in the arrival of new foodstuffs from the Americas.[5] The era of Portuguese commercial dominance would eventually be replaced by that of their northern European rivals, the Dutch, the English, and the Danes, in the seventeenth century. New

[3] Roxanne Tremblay, Manuelle Landry-Cuerrier, and Murray Humphries, 'Culture and the Social-Ecology of Local Food Use by Indigenous Communities in Northern North America', *Ecology and Society* 25, no. 2 (2020).

[4] Douglas, *Purity*; Douglas, *Natural Symbols*; Claude Fischler, 'Food, Self and Identity', *Social Science Information* 27, no. 2 (1988): 284–85.

[5] McCann, *Maize and Grace*; La Fleur, *Fusion Foodways*.

foodstuffs brought by the Portuguese to provision their trade outposts would transform the foodways, culture, and the historical trajectory of local polities. This chapter then seeks to contextualize traditional foods and how local people came to have a long-term biological relationship with New World foods.

Ghanaian Geography

Present-day Ghana is roughly the size of the United Kingdom. Like the Italian peninsula, whose northern region is different from its southern region in terms of geography, language, and cuisine, Ghana's geographies similarly feature diverse languages and distinct staple foods. Each cultural region corresponded roughly with a distinctive geography. This chapter broadly outlines the major ecological differences and the corresponding societies and states in Ghana: the northern savanna zone, the forested centre, and the coast. The term eco-culinary zone is being applied to each of these regions in order to illustrate the differences in the localised ecology. It is worth noting here that Ghanaian historiography tended to focus on the coast and on the Akan forest states, especially Asante. This tendency corresponds to the order in which Europeans met local people and traded with them on the coast, and with the perceived threat that the powerful Akan states presented to local peoples and Europeans.[6] The Ewe on the east side of the Volta River, for example, did not properly come under British influence until German-ruled Togoland was divided between Britain and France in 1914.[7] Written sources for the Gold Coast, based upon the accounts of traders and travellers, go back to the fifteenth century, as compared with those of the Northern Region which do not properly begin until the late nineteenth century. Furthermore, the anthropological focus of the scholarship in the Northern Region meant that specific ethnic groups like the Gonja or the Tallensi, who have been

[6] Reindorf, *History*, 111–28.

[7] Akosua Anyidoho and M. E. Kropp Dakubu, 'Ghana: Indigenous Languages, English, and an Emerging National Identity', in *Language and National Identity in Africa*, ed. Andrew Simpson (Oxford: Oxford University Press, 2008), 153.

the focus of various ethnographies, will broadly represent the foodways of that region.[8]

Ecological Zones

Ghana has a tropical climate, characterised by a seasonal change is from the wet season to the dry season. Northern Ghana lies in the savanna and was divided into five administrative units: Upper West, Upper East, Northern, Savanna and North East regions (in 2019 the Northern region was divided into three new regions: Savanna, Northern and North East). This area is drained by the Lower Black Volta River and its tributaries the White and Red Voltas, as well as the Oti and Daka Rivers. It is bounded by Burkina Faso in the north, Togo to the east, Côte d'Ivoire in the west, and Brong Ahafo Region to the south. The northern part of the country has just one rainy season from roughly August to September/October. Rainfall averaged from between 76 and 101 centimetres. The north constitutes more than half of Ghana and is composed of savanna woodland with high grass and semi deciduous trees such as the shea tree or the thorny acacia tree and shorter grasses further northeast. Cash crops from this area include shea butter, groundnuts (peanuts), and more recently, cotton, and tomatoes.[9] Food crops in this region include millet and sorghum. Yams and maize are found in areas with more rainfall. Rice is grown in flooded areas drained by the Volta River system. Cattle raising is more common here than in the south, as the tsetse fly is less prevalent.[10] Northern Ghana presents special challenges to the food producer and cook due to its single rainy season.

---

[8] For more on how the work of ethnographers of the Northern Region have structured the scholarship of that area, see the introduction of: Jean Allman and John Parker, *Tongnaab: The History of a West African God* (Bloomington: Indiana University Press, 2005).

[9] Benjamin Asinyo, Charles Frimpong, and Emmanuel Amankwah, 'Original Article: The State of Cotton Production in Northern Ghana', *International Journal of Fiber and Textile Research*, 2015, 7; Martha A. Awo, 'Marketing and Market Queens: A Case of Tomato Farmers in the Upper East Region of Ghana' (PhD diss., Rheinischen Friedrich-Wilhelms-Universität zu Bonn, 2010).

[10] Cardinall, *Natives of the Northern Territories*, 83; Fortes and Fortes, 'Food in the Domestic Economy', 42; Rattray, *Tribes*, 296, 318, 360, 369, 516; Jack Goody, *Cooking*, 51–69.

By contrast, the southern half of the country has two distinct rainy seasons. The first is roughly from March to July and the second from September until November. Within the southern half of the country, the central forested region contains evergreen and deciduous trees up to 60 m high. It typically features dense undergrowth with hills and mountains, and is well watered by the Pra, Birim, Ankobra, and Tano rivers. Lake Bosomtwi is a significant (and sacred) body of water. Besides all these riches, the forested region receives on average up to 203 centimetres of rain per year.[11] The number one export of this zone is currently gold, followed by cocoa. Additional cash crops include kola nuts traded to the Sudanic zone. Oil palms and coconut palms grow best in a narrow belt closer to the coast in present day Western, Central and Eastern administrative regions of southern Ghana.[12] Food crops in the rainforest tend to be root crops: yam, cocoyam, cassava, and Asian yam. Maize, rice (in flooded areas), bananas, and plantains also grow well here.[13] The central forested region receives the most rainfall, producing much of the produce exported to global markets.

The area up to 60 miles from the ocean is considered the coastal zone. Composed of level, grassy plains with small ranges of hills or peaks, it has few trees. Annual rainfall is on average half that of the central forested zone.[14] Scrub and grassland stretch from the Volta Delta to the Pra River. Fishing and salt making are the two main productive activities. The biggest locally consumed export of this region is fish caught off the coast. In the mid-1990s 360,000 tonnes came from these waters.[15] That number

[11] Colin G. Wise, 'Climatic Anomalies on the Accra Plain', *Geography* 29, no. 2 (1944): 35–38.

[12] See: Kwabena Ofosu-Budu and Daniel Bruce Sarpong, 'Oil Palm Industry Growth in Africa: A Value Chain and Smallholders' Study for Ghana', in *Rebuilding West Africa's Food Potential*, ed. A. Elbehri (Rome: FAO/IFAD, n.d.), 44; Donna J. E. Maier, 'Precolonial Palm Oil Production And Gender Division Of Labor In Nineteenth-Century Gold Coast And Togoland', *African Economic History* 37 (2009): 1–32; Martin Lynn, 'The Profitability of the Early Nineteenth-Century Palm Oil Trade', *African Economic History*, no. 20 (1992): 77–97.

[13] T. Edward Bowdich, *Mission from Cape Coast Castle to Ashantee* (London: J. Murray, 1819), 319; Edmond Perregaux, 'Chez Les Achanti', *Société neuchâteloise de géographie* 17 (1906): 19; Huydecoper, *Diary*, 16.

[14] Wise, 'Climatic Anomalies'; Peter Kwabenah Acheampong, 'Rainfall Anomaly along the Coast of Ghana', *Geografiska Annaler* 64, no. 3/4 (1982): 199–211.

[15] J. Atta-Quayson, ed., *Macmillan Atlas for Ghana* (London: Macmillan Education Ltd., 1995).

increased to 451,227 tonnes in 2001.[16] The most prevalent traditional food crop was millet.[17] However, by the early nineteenth century cassava intercropped with maize became the main starches, and okro (okra), peppers, tomatoes and garden eggs (aubergine) were also foundational to the foodways there.[18] It is the coastal areas that historically interacted with European traders, which created a unique urbanized and cosmopolitan social environment.

Ethnicity and Language

Within these eco-culinary zones exist many ethnic groups, and these groups used both language and food as markers of ethnic identity. Ghana is a multicultural country with more than 60 indigenous languages, two-thirds of which are spoken in northern Ghana. Most Ghanaians are bi-lingual or multi-lingual.[19] There are two main language families in Ghana, the Gur and the Kwa, both branches of the Niger-Congo phylum, with two languages—Ligby and Bisa—belonging to the Mande phylum. In the north the Gur language family is again subdivided into the Oti-Volta group and the Grusi.[20] In the south of the country the most prevalent languages are of the Kwa group of which Twi serves as the lingua franca, while English serves as a trade language in the north. Hausa, a Chadic language spoken in the north, rounds out the language offerings.

[16] 'FAO Fishery Country Profile', The Republic of Ghana, General Economic Data—March 2004, accessed 20 October 2019, http://www.fao.org/fi/oldsite/FCP/en/GHA/profile.htm.

[17] De Marées, *Description*, 113; Adam Jones, ed., *German Sources for West African History, 1599–1669* (Wiesbaden: Steiner, 1983), 207–09; Bosman, *Description*, 391.

[18] Henry Meredith, *An Account of the Gold Coast of Africa* (London: Longman, Hurst, Rees, Orme, and Brown, 1812), 179, 213, 223; De Marees, *Description*, 62; Paul Erdmann Isert, *Letters on West Africa and the Slave Trade*, trans. Selena Axelrod Winsnes (Legon: Sub-Saharan Publishers, 2007), 168–70.

[19] Sebastian Bemile, 'Promotion of Ghanaian Languages and Its Impact on National Unity: The Dagara Language Case', in *Ethnicity in Ghana: The Limits of Invention*, ed. Carola Lentz and Paul Nugent (New York: Palgrave USA, 2000), 204, 211.

[20] Tony Naden, 'The Gur Languages', in *The Languages of Ghana*, ed. M. E, Kropp Dakubu (London: International African Institute, 1988), 12–49.

The South

Of the Kwa languages in the south, Akan is spoken by the most people. According to the 2020 census, the current population is around 31 million, of whom an estimated 18 million speak Twi as a first or second language. This is followed by Ewe speakers who make up 12% of the total population.[21] Akan languages are commonly known as Twi and are spoken by eleven subgroups, most of whom live in the forested regions. The largest of these eleven are the Asante, Fante, Akyem, Akwamu, and Nzima.[22]

The story of the Akan ethno-linguistic group is one of migration, expansion, and consolidation. Akan traditions indicate that they arrived in Ghana from the north, perhaps between the eleventh to the thirteenth century, where they founded Bono-Mansu (near present-day Techiman) and traded gold.[23] By the mid-seventeenth century, two Akan kingdoms, Denkyira and Akwamu, had risen to a position of dominance in the central forest zone and in the hinterland of the eastern Gold Coast. Denkyira was defeated by the Kumasihene (Chief of Kumasi) Osei Tutu in 1701 and became a vassal state of the new Asante kingdom under his leadership.[24]

Fante oral traditions suggest that they migrated to the coast from Bono-Mansu at the beginning of the seventeenth century and again in another wave after the defeat of Bono-Mansu at the hands of the Asante around 1740. They too organised themselves into a confederacy.[25] Other Akan states located in the eastern forest, rose and fell to eventually become vassals of the expanding Asante. Among these were the Akwamu and the

[21] Anyidoho and Kropp Dakubu, 'Ghana Languages', 142–51; 'Ghana Population (2020),' Elaboration of data by United Nations, Worldometer, last modified 1 July 2020, https://www.worldometers.info/world-population/ghana-population/#:~:text=Ghana%202020%20population%20is%20estimated,of%20the%20total%20world%20population.

[22] Gocking, *History*, 8.

[23] Rattray, *Ashanti*; Bosman, *Description*; Bowdich, *Mission*; Eva L. R. Meyerowitz, *Akan Traditions of Origin* (London: Faber and Faber, 1952).

[24] Bosman, *Description*, 76; T. C. McCaskie, 'Denkyira in the Making of Asante c. 1660–1720', *Journal of African History* 48, no. 1 (2007): 1–25; Rattray, *Ashanti*, 288–91.

[25] Madeline Manoukian, *Akan and Ga-Adangme Peoples of the Gold Coast* (London: Oxford University Press, 1950), 9–15; Rattray, *Ashanti*, 113.

Akim. Under Asantehene Osei Tutu and his successor Opoku Ware, many of the former Akan kingdoms were incorporated as Asante tributary states from Banda in the west to Krakye on the Volta River in the east.

Akan speakers tended to be matrilineal and formed into *abusua* (female lineages) where property and titles passed from uncle to nephew. Abusua shared corporate responsibility for debts, crimes, and funeral expenses.[26] *Asafo* membership (military associations) had a tendency to be patrilineal among this ethno-linguistic group.[27] The Asante were organised into eight clans and the Fante into seven clans, and many of the other Akan speaking groups were organised into the same clans in a similar fashion. A clan is a lineage group descended from the same ancestress.[28] Each clan was customarily associated with certain localities and legends of migration. The most important clan was the Asantehene's or king's clan—Oyoko.

Skilled colonisers, the Fante acculturated the native Guan-speaking peoples of the coastal region in the eighteenth century.[29] In the migration story of the leading Aboradze clan, a hunter was seeking new lands. He grew hungry. Just then, he encountered a lion who, instead of eating him, instructed him to place a piece of dry wood in his teeth and to rapidly pull it out. This created sparks, thus enabling the hunter to make fire. Upon that fire he roasted plantain. He tasted it and saw it was good.[30] This story illustrated the importance of plantain as the main starchy food of the leading Aboradze clan. The Fante name for plantain, *brodze*, is said to mean "over all others" and reflected the ecological makeup of Fante land—from west of Accra to Sekondi-Takoradi—which was forested and amenable to plantain cultivation.[31] The name of the leading Aboradze clan bore similarities to the Akan name for the stump of a plantain tree

---

[26] Manoukian, *Akan and Ga*, 22.

[27] Gocking, *History*, 8.

[28] Ivor Wilks, 'The State of the Akan and the Akan States: A Discursion', *Cahiers d'études africaines* 22, no. 87 (1982): 231–49.

[29] Mary McCarthy, *Social Change and the Growth of British Power in the Gold Coast: The Fante States, 1807–1874* (Lanham: University Press of America, 1983), 11.

[30] G. A. Acquah, *The Fantse of Ghana: A History* (Hull: University, 1957), 11.

[31] Ibid.

from which new trees grow—*abrode sé*.[32] Additionally, the word *brod* meant loaf in both Danish ("*brød*") and in Akan.[33] This evidence indicated that from the time of their flight to the coast in the eighteenth century, the Fante considered plantain to be their principal starch.

The Fante's geographical location in a partially forested coastal environment meant in addition to plantain, they were also particularly suited to produce millet, yam, and later, cassava. Millet, both white and red, was boiled to the consistency of bread (a stiff porridge), and yams were boiled or roasted. These starches were accompanied by palm oil and dried fish.[34] The Fante produced and ate "Portuguese maize" and rice that was grown in nearby Axim.[35] Their specialisation in ocean fishing was an important source of nutrition and revenue derived from trade with the interior and from Europeans on the coast.[36] Komenda was the centre of production for palm oil, and the Fante also produced salt, palm wine, and palm oil.[37] The Fante's early contact with Europeans in the Atlantic trade as provisioners of slave ships and European forts, and their role as slave traders, gave them access to a variety of foods, tableware, and household goods on a different scale from some of the other groups on the coast.

As merchants and middlemen, the Fante were accustomed to consuming luxury imported goods as they controlled the main trading centres of the Gold Coast during the Atlantic era. Europeans often married into Fante families, which had the effect of facilitating trade relationships. Bosman characterised the Fante in the eighteenth century as grain sellers, highly productive fishermen, and merchants.[38] In 1728, some of the goods the Fante consumed included the following items the Dutch stocked in their trading forts: garlic, tea, coffee, chocolate, preserves, candied fruits, bottled beer, butter, biscuits, cheese, pickled herring, barrels

[32] J. G. Christaller, *A Dictionary of the Asante and Fante Language Called Tshi* (Basel: Evangelical Missionary Society, 1881), 574.

[33] La Fleur, *Fusion Foodways*, 118.

[34] Rask, *Two Views*, 124, 286–99; Bosman, *Description*, 297.

[35] Ibid., 297–98.

[36] Bosman, 58, 124.

[37] Rask, *Two Views*, 56.

[38] Bosman, *Description*, 57–58.

of salt pork and beef—all used as gifts to lubricate the wheels of trade.[39] The trappings of rank and wealth such as silverware, glassware, china, damask napkins, brass pans, and silk damask tents with poles were distributed to both leading traders and chiefs.[40] The Braffo or chief, governed a tiered, decentralised organisation that was designed to redistribute the tribute received from European traders. This organisation created an upward and downward flow of wealth and enabled the Fante to develop into a consumer society of foreign goods.[41]

The Gã and Adangme (or Dangme) situation differed from the Akan significantly in that they were a much smaller subsection of the Kwa ethno-linguistic group. However, they tended to be influential due to their geographical location on the coast. The Gã migrated from the east by the late sixteenth century into the La Adangme plains, which were inhabited by the *Kpesi* people. The Adangme split off and settled the coastal area from Kpone to Ada on the Volta, and as far into the hinterland as Manya Krobo. The Shai, Ningo, and Krobo people are all Dangme speakers.[42]

According to a 2011 population study, Gã-Dangme ethnic group makes up only 8 percent of the total population of Ghana.[43] The Gã live mostly in and around Accra, the nation's capital, of which they make up 29 percent of the population.[44] The maxim *Gã se gbe ji gbe* means "All roads lead to Accra", indicates the town's prominence and the occupants' conception of themselves as leading people.[45] Accra has been a trading centre since the seventeenth century and became the capital of the British

---

[39] Stanley B. Alpern, 'What Africans Got for Their Slaves: A Master List of European Trade Goods', *History in Africa* 22 (1995): 28.

[40] Alpern, 29.

[41] McCarthy, *Social Change*, 11, 13; Bosman, *Description*; Henry Meredith, *Account*; Ty M. Reese, '"Eating" Luxury: Fante Middlemen, British Goods, and Changing Dependencies on the Gold Coast, 1750–1821', *The William and Mary Quarterly* 66, no. 4 (2009): 851–72.

[42] Manoukian, *Akan and Ga*, 66, 105; M. J. Field, 'The Agricultural System of the Manya-Krobo of the Gold Coast', *Africa* 14, no. 2 (1943): 60.

[43] George Owusu and Samuel Agyei-Mensah, 'A Comparative Study of Ethnic Residential Segregation in Ghana's Two Largest Cities, Accra and Kumasi', *Population and Environment* 32, no. 4 (1 June 2011): 343.

[44] Ibid.

[45] John Parker, *Making the Town: Ga State and Society in Early Colonial Accra* (Oxford: James Currey, 2000), viiii.

Crown Colony of the Gold Coast in 1877.[46] The Gã in Accra came under Akan-speaking Akwamu rule between 1681 and 1730, and Gã country was under Asante rule from 1742 until the mid-1820s.[47] Unlike members of the Akan ethnic group, the Gã are patrilineal but are greatly influenced by Akan language and culture, adopting elements of matrilineal inheritance and borrowing vocabulary from the Akan Twi language. Gã territory is roughly bounded by the Gulf of Guinea in the south, the Akuapem escarpment in the north, and the Sakumo River and lagoon in the West. The Laloi Lagoon forms the eastern border, which merges with the Dangme-speaking peoples, who occupy the rest of the coast down to the Volta River.[48]

The Gã formed into independent towns or *man* for mutual protection from slave raiding at the end of the seventeenth century.[49] Towns were centres of military companies named *asafoi* (sing. *asafo*), which were organised by *akutsei* (sing. *akutso*) or quarters and by age grades/ranges.[50] These quarters were composed of colonists who attached themselves to an original group of settlers, the *Kpesi,* preserving their own culture but recognising the supremacy of the senior god of the town and its priest. The *wulomo* (pl. *wulomei*) or priest of the senior *Kpesi* god of each town was recognised as the town landlord or *sitse*.[51] At the end of the seventeenth century, the Gã formed into seven different towns along the coast: Nleshi, Kinka, Osu, La, Teshi, Nungwa, and Tema. Three of these—Nleshi, Kinka, and Osu—drew together in the early colonial period to become Accra.[52] Each town colonised a hinterland of farming hamlets stretching back between 20–70 miles from the coast, populated by both free and unfree farmers who were considered part of the town. There are also several miles of fishing villages strung along the coast from east to west that

[46] Anyidoho and Kropp Dakubu, 'Ghana Languages', 154; Parker, *Making the Town*, xvii.

[47] See: John Kofi Fynn, *Asante and Its Neighbours, 1700–1807* (Evanston: Northwestern University Press, 1971); Ray A. Kea, 'Trade, State Formation and Warfare on the Gold Coast, 1600–1826' (unpublished PhD diss., University of London, 1974), 196–99.

[48] Parker, *Making the Town*, 2.

[49] Gocking, *History*, 10; Manoukian, *Akan and Ga*, 69, 73; Zimmermann, *Grammatical Sketch*, 190–93.

[50] Parker, *Making the Town*, 19.

[51] Gocking, *History*, 10; Manoukian, *Akan and Ga*, 67, 81.

[52] Parker, *Making the Town*, 2.

were part of Gã territory.[53] Both farming and fishing were seasonal in Accra: "In the months of Jan., Feb., and March there are usually, at Akra on the Gold Coast, two varieties of small fish. The first type is called sardines, which have a lot of small bones. It can be salted and is used like herring, which it closely resembles."[54] Sardines were among the most popular types of fish, followed by *cinqesous* (sea bream), which were both fished in the summer months.

From its sixteenth century Portuguese introduction to the coast, maize was grown in the Akuapem Hills (dubbed the "breadbasket of Accra") just northeast of Accra, and was traded in Accra for salt and dried fish.[55] Maize easily lent itself to contemporary processing methods for both sorghum and millet into the local bread—*kenkey*, which became a staple and prepared market food as early as 1600.[56] *Kenkey* was processed by grinding maize on an inclined stone while continuously dampening with water until a fine dough was formed. This dough was then allowed to sour and baked in small ovens.[57] By the nineteenth century maize became a commercial crop used to supply growing urban markets:

From the seventeenth century, noblemen diversified their trading affairs to include establishing plantations of food crops—especially maize. They established these slave plantations in order to exploit the exploding demand for foodstuffs presented by growing coastal towns populated by non-farming traders, soldiers, and craftspeople to whom they could trade foodstuffs against imported commercial goods.[58] Maize became the starch or bread called *kenkey* that formed the foundation of the Gã main meal of the day, effectively displacing millet and sorghum beginning in the seventeenth century. Maize also became the foundation of the local economy and the *raison d'être* for participation in the global economy for many coastal peoples.

Much like other ethnic groups on the coast, the Ewe history was a tale of exodus. This exodus began in Benin in the late sixteenth century,

---

53 Manoukian, *Akan and Ga*, 66–67; Parker, *Making the Town*, 7.

54 Rask, *Two Views*, 40.

55 Meredith, *Account*, 226–27.

56 De Marées, *Description*, 272.

57 Isert, *Letters*, 167.

58 La Fleur, *Fusion Foodways*, 132.

through present day Notsie in Togo, to south-eastern Ghana. In towns physically separated from the mainland (and its potential for violence), they lived on the other side of the lagoons and mangrove swamps of the coast. Sandy soils were not capable of supporting intensified traditional millet production and caused the Ewe to turn to cassava, which they named *agbeli*, or "there is life".[59] The lagoon, riverain, and ocean environment of the Ewe on the eastern Gold Coast necessitated a deliberate approach to their provisioning strategies once they migrated to the region. They adjusted to lagoon fishing, salt making, and cattle and fowl raising for sale to the western Gold Coast, as well as the manufacturing of baskets and cloth. Their late seventeenth- and eighteenth-century economy was dominated by the trade of surplus dried fish and salt inland, which was exchanged for millet, rice, and palm oil. Ewe social organisation at this time emphasized endogamous marriages to protect access to production land from immigrants looking to escape the expanding slave kingdoms of Asante in the west and Dahomey in the east.[60]

Beginning in 1679, the coastal Ewe received the first wave of what would be successive migrations of refugees fleeing conflicts from the expanding Akwamu state on the west side of the Volta River.[61] Ewe speakers became early adopters of cassava as a foodstuff to cope with the waves of refugees escaping the slave trade. The Ewe developed new ways to prepare cassava which was poisonous and required soaking before cooking to make it safe to eat. Initially, cassava was prepared in the coastal region by simply uprooting and boiling the roots. The Ewe then expanded the culinary cassava repertoire by creating new ways to prepare their starchy cassava staple dough *agbelima* (*agbeli* + *ma*, "dough"), which they fermented to improve the taste and nutrition of the starch. Other dishes included *agbeli kaklo* (fresh grated cassava dough salted and formed into balls and fried) and *yakayake* (dried and salted cassava, shaped in balls and steamed). *Akple* was a stiff, fermented cassava porridge or paste that

[59] Francis Agbodeka, ed., *Handbook of Eweland: The Ewes of Southeastern Ghana*, vol. 1 (Accra: Woeli Publishing Services, 1997), 197; La Fleur, *Fusion Foodways*, 167.

[60] Emmanuel Akyeampong, *Between the Sea and the Lagoon: An Eco-Social History of the Anlo of Southeastern Ghana: C. 1850 to Recent Times* (Athens: Ohio University Press, 2001), 37–38.

[61] Sandra E. Greene, *Gender, Ethnicity, and Social Change on the Upper Slave Coast* (Portsmouth: Heinemann, 1996), 20–26, 68–71.

was the accompaniment of soups prepared with vegetables, fish, and meat.[62] One of the biggest changes in Ewe foodways in the eighteenth century then, was the transition from millet used to prepare bread as their starchy staple, to cassava.[63] Cassava's position as the most widely used root crop and main starchy staple in Ewe cuisine, meant that, in time, cassava became a marker of general Ewe ethnic identity.

Initially, those people fleeing from the east settled in Notsie, but due to the cruelty of King Agorkorli, some decided to flee further west in the early seventeenth century. One of the first places they rested was at Tsevie in present day Togo, where they attempted to grow provisions. The significance of the meaning from which Tsevie takes its name—*tse* meaning "yield" and *vi* meaning "little"—came after wild pigs ate most of the cowpea (black-eyed pea) crop. Some of the refugees wanted to remain to allow the cowpeas to "grow a little more" or to "let it yield small" before leaving the plantation.[64] Others wanted to press on, causing the group to split. Some continued south to the coast and others stayed behind at Tsevie plantation, as it is now called. This event is commemorated each year at the *Ayiza* or *Hogbetsoso* harvest festival on the coast where the ceremonial food is the cowpea.[65] *Hogbetsoso* is from the Ewe *Hogbesotsoza*—*hogbe* (meaning homeland, in this case Notsie), *tsotso* (meaning exodus), and *za* (festival).[66] Bosman catalogued no less than four bean varieties consumed by the Ewe in the eighteenth century. Some were runner beans and others were subterranean.[67] Both the major festival commemorating their flight to safety, as well as the commentary of contemporary observers tell us that the Ewe relied heavily upon beans as a major part of their survival strategy during the seventeenth century. Beans, like cassava, became a marker of general Ewe ethnic identity.

---

[62] E. V. Doku, *Cassava in Ghana* (Accra: Ghana University Press, 1969), 32–34; Agbodeka, *Handbook of Eweland*, 1:198.

[63] Bosman, *Description*, 289; Rask, *Two Views*, 138.

[64] A. B. Ellis, *The Ewe-Speaking Peoples of the Slave Coast of West Africa* (London: Chapman and Hall, 1890), 215.

[65] Agbodeka, *Handbook of Eweland*.

[66] Meera Venkatachalam, *Slavery, Memory and Religion in Southeastern Ghana, c.1850–Present* (Cambridge: Cambridge University Press, 2015), 152.

[67] Bosman, *Description*, 300–301.

Continued instability in the eighteenth and nineteenth centuries reinforced the reliance upon beans and cassava in the Ewe diet. The Ewe located on either side of the Volta river share a language and history with Ewe speakers on the Togo side of the border. In the eighteenth and nineteenth centuries, the Asante in the west and the Dahomey in the east drove many peoples, including the Ewe, into the Togoland Hills, a range which runs in a north-easterly direction and forms the international border between Ghana and Togo. As in the Fante region, the autochthonous people were mostly absorbed by various migrants like the Ewe, but still retain a sense of their own culture and language. This resulted in much mixing of both cultural and political elements. Some Ewe went to the Togoland Hills, some were captured and taken to Kumasi, some attempted to return to their homes but ended up settling in new locations.[68]

Ewe became a sort of *lingua franca* for the Togoland Hills as the area was already settled by many non-Ewe speakers such as the Lolobi and Likpe. A mixture of patrilineal and matrilineal social organisation existed, as this area too was heavily influenced by their Akan neighbours.[69] An example of an Ewe settlement was the polity of Anlo on the Atlantic littoral in the south-eastern corner of Ghana. Organised into clans, or *hlowo*, composed of traders, conquerors, slaves, and refugees who settled among the locals beginning in 1679, the clans were patrilineal military coalitions within which ethnic insiders and newcomers were assimilated. The gods of the oldest known indigenous group were given special reverence. Millet and sorghum were cultivated by women, who traditionally bestowed plots on their daughters. Anlo's location on a narrow sandspit between the Atlantic and the Keta Lagoon made lagoon fishing an important industry in which women participated by collecting shellfish and

[68] Paul Nugent, 'A Regional Melting Pot: The Ewe and Their Neighbours in Ghana-Togo', in *The Ewe of Togo and Benin*, ed. Benjamin Nicholas Lawrance (Accra: Woeli Publishing Services, 2005), 31.

[69] Gocking, *The History of Ghana*, 9–10; Anyidoho and Kropp Dakubu, 'Ghana Languages', 153; Paul Nugent, '"A Few Lesser Peoples": The Central Togo Minorities and Their Ewe Neighbours', in *Ethnicity in Ghana: The Limits of Invention*, ed. Carola Lentz and Paul Nugent (New York: Palgrave USA, 2000); M. E. Kropp Dakubu and K. C. Ford, 'The Central Togo Languages', in *The Languages of Ghana* (New York: International African Institute, 1988), 125.

smoking, drying, and salting the catch of their husbands and/or male relatives.[70]

The North

Much like the coastal Fante, Gã and Ewe, the northern savanna regions of Ghana were populated by indigenous people interspersed with those who migrated and settled among them. The Mossi invaded in the early fifteenth century and, imposing their rule over the local people using cavalry technology, founded the kingdoms of Mamprugu and Dagbon.[71] Descendants of the Mossi were to found in Wa, in the northwest, and in Nunumba, on the eastern frontier of Ghana. Mande-speaking people also entered the region and formed the kingdom of Gonja on the Black Volta in the mid-sixteenth century. Generally, these invaders intermarried with locals and adopted their languages and most social norms.[72] The social organisation of the region was therefore composed of a mixture of centralised states and small non-centralized clans.[73]

There was great linguistic diversity in the Northern Region, but the numerous related languages are part of the Oti-Volta language family, many of which are mutually intelligible. Two of the most widespread Niger-Congo languages are Dagaare and Dagbani, Oti-Volta languages within the Gur branch of the Niger-Congo language family. Not only Gur languages are spoken in the northern Ghana, however; a Kwa language is spoken by the Gonja in the southeast, and in the northeast corner of Ghana (Chokosi). Both these Kwa languages are closer to Akan in the south than any of the languages in northern Ghana.[74]

---

[70] Greene, *Gender*, 1–5.

[71] Jack Goody, *Technology, Tradition, and the State in Africa* (London: Oxford University Press, 1971).

[72] Allman and Parker, *Tongnaab*, 25; Louis Tauxier, *Le noir de Bondoukou* (Paris: E. Leroux, 1921); Rattray, *Tribes*.

[73] A K Awedoba, 'The Peoples of Northern Ghana', National Commission on Culture Website—www.ghanaculture.gov.gh, 2006, http://lagim.blogs.brynmawr.edu/files/2015/03/The-Peoples-of-Northern-Ghana.pdf.; Gocking, *History*.

[74] Awedoba, 'The Peoples of Northern Ghana'; Anyidoho and Kropp Dakubu, 'Ghana Languages', 155.

Some ethnic groups in this area were united by shared myths, which established a common ancestor, and there was a tendency to patrilineal inheritance. For those with kingships, the leadership tended to alternate between gates (lineages). Northerners traditionally farmed millets, sorghum, legumes, groundnuts, livestock, and—where soil and climate permitted—maize. Many more migrated to the south for work, where they often suffered from low prestige due to southern Ghanaians perception of them as mere domestic and manual labourers.[75]

New Foods

The sixteenth century was a time of widespread experimentation on the Gold Coast and its hinterland as farmers added new starchy staples arriving from the Americas and elsewhere in Africa to their culinary repertoire. Two of those new foods already mentioned in the previous brief descriptions of microclimates, were plantains and maize. Maize originated in Meso-America and arrived via the Columbian Exchange. Local farmers did not just adopt maize wholesale, but selectively, gradually, and cautiously, based upon local tastes.[76]

Five major maize types were propagated in the New World: pop, floury, flint, dent, and sweet. Flint and floury were those which adapted well to West African micro ecologies. Flint originated in Caribbean biomes and acclimated well to areas where sorghum and millet thrived. It was especially useful as a stopgap for the hungry season before the slower growing sorghum was mature. Flint maize was composed of hard, slightly translucent grains, with harder starches in its endosperm, and could be prepared like couscous or rice. Because each kernel had a tough outer layer to protect the soft endosperm, it is likened to being hard as flint. The Spanish encountered flint in the Caribbean—a brightly coloured (mostly red), hard starch with an early maturation period. Flour or floury maize has a soft starchy endosperm and a thin pericarp. It is primarily used to make corn flour. Floury maize was very adaptable to the forest-savanna mosaic.

[75] Anyidoho and Kropp Dakubu, 'Ghana Languages', 157.

[76] La Fleur, *Fusion Foodways*, 55; McCann, *Maize and Grace*, 23.

It was usually hand milled and transformed into kenkey—the stiff porridge steamed in its husk and eaten with palm oil and vegetable or bush meat stews. It would come to replace flint types adopted during the slave trade.[77]

Both McCann in *Maize and Grace* and La Fleur in *Fusion Foodways* noted the debate from the mid-1950s about who introduced different varieties of maize to West Africa and when they were introduced. The account of Pieter De Marees published in 1602 is the first primary source which attests with certainty to the presence of maize on the West African coast:

> This Corn is called Maize by the Indians; the Portuguese or Spanish call it Indian Wheat and the Italians Corn of Turkey. It is a grain which is known virtually throughout the world, and was brought from the West Indies to São Tomé. The [Portuguese] people of São Tomé brought it here (after establishing their strongholds and Castles) to supply their needs, and sowed it here in these Lands; for before the arrival of the Portuguese the inhabitants of these Countries did not know it. Thus the Portuguese sowed it there and distributed it amongst the Savages, so that the country is now full of it and at present it grows in abundance. They use it in a mixture, grinding it together with their Millie, sometimes taking half Millie [sorghum or millet] and half Maize.[78]

De Marees then attests that the Portuguese brought maize to the Gold Coast via the Caribbean to São Tomé, and then to the Gold Coast. Another early source that mentioned maize on the Gold Coast is that of Olfert Dapper published in1668:

> Their main food is bread, called kantjens [*kenkey*] by them, fried or cooked from millie or Mais, which is covered with some palm oil, and sometimes dipped in green herb(s), and prepared in this way: the millie is first pounded in bits with a stick in a stone mortar, which they call threshing, afterwards it is winnowed clean in a wooden dish: the women then rub this threshed millie, (something they do twice every day,) on a flat stone, standing a

[77] McCann, *Maize and Grace*, 49.

[78] De Marées, *Description*, 113.

man's length from the earth, with another stone a foot long, with her hands to flour, in the same way as painters usually rub their verruwen [?], which is not a small labour for her, though the men look down on it. The millie, ground to flour, do they mix with water, and form into cakes or balls, the size of a couple of fists, which they bake for a short time on a hot floor, or cook, wound in towels or cloths. Some mix the millie with half Mays (or Turkse tarwe), others also use only Mays and bake bread from that.[79]

In the same account, Dapper also asserted that maize was brought to the west coast of Africa by the Portuguese via the island of São Tomé. Both these accounts demonstrated that maize was combined with native millet and/or sorghum before cooking and eating in the form of *kenkey*. This would indicate that by the mid-seventeenth century, maize had not yet completely supplanted millet and sorghum on the coast but was commonly used to supplement other crops.

Since the 1950s there has been debate as to when different types of maize—flint versus floury maize—were introduced to West Africa. In 1955 Roland Portères used linguistic evidence to determine that flint varieties arrived from Egypt via the Mediterranean after 1517. In the sixteenth century, maize was variously known as *masa/masara/masar/mahar* in Northern Nigeria. Portères maintained that the evidence for an Egyptian origin of maize lay in the Nigerian name for maize: "However, the term Masar, common to all these various names for corn [maize], is none other than the very name of Egypt in Arabic and in the North Nigerian dialects."[80] In other words, flint maize was called "Egypt" in Northern Nigerian languages and this Portères believed proved its Egyptian origins.

In his 1962 article, Frank Willett's objective was to disprove the theory that maize was introduced into Africa by Arabs.[81] During his research, he observed that Porteres's distinction between floury and flint maize in

[79] Olfert Dapper, *Naukeurige beschrijvinge der Afrikaensche gewesten van Egypten, Barbaryen, Libyen, Biledulgerid, Negroslant, Guinea, Ethiopiën, Abyssinie* (Amsterdam: Jacob van Meurs, 1668), 463.

[80] Roland Portères, 'L'Introduction du Maïs en Afrique', *Journal d'agriculture traditionnelle* 2, no. 5 (1955): 223.

[81] M. D. W. Jeffreys, 'Pre-Columbian Maize in Africa', *Nature* 172, no. 4386 (November 1953): 965–66.; Frank Willett, 'The Introduction of Maize into West Africa: An Assessment of Recent Evidence', *Africa* 32, no. 1 (1962): 1–13.

West Africa introduced new information for dating the arrival of maize. He noted the current preponderance of floury maize on the Guinea Coast and, knowing that flint types of maize did not reach Egypt until about 1520, Willet estimated the earliest possible date for the meeting of the two types to be from approximately 1530 to 1540. By Willett's estimate, flint maize was introduced after 1520 and the floury variety—which originated in Brazil, Mexico and the Andes—from about 1525 to 1535, possibly via São Tomé.[82] The timing of the arrival of floury maize from Iberian American colonies proved its Portuguese introduction to West Africa.

LaFleur in *Fusion Foodways* contended that identifying precise dates for these early experiments with maize in the Gold Coast is, as with all other agricultural innovations, impossible, but the earliest maize in this area and in the rest of Africa and Eurasia, was certainly post-Columbian and introduced by the Portuguese on the coast. Fante-speaking farmers near Elmina were likely the first Gold Coasters to experiment with maize and welcomed this crop as a cereal that was an abundant producer. Linguistic evidence showed the Fante drew a relationship between their native millet (*wi*—'millet') and maize from overseas when it was named *oborowi*. Thus, the commonly held assertion that floury maize was introduced on the coast by the Portuguese from their American colonies for the feeding of slaves was correct, mainly because there existed no linguistic evidence of maize cultivation before the sixteenth century.

However it was introduced, by the late seventeenth century maize had largely replaced both millet and sorghum in many coastal areas of West Africa.[83] As well as the Fante, the Banda, the Gã and the Guang speakers north of the Volta River were relatively early adopters. Two hybrid varieties were developed that were particularly adapted to local environmental conditions: high yielding maize grown during the longer rainy season from April to June and a shorter variety grown during the shorter rainy season from September to November. From these types of yields, hard,

[82] Willett, 'The Introduction of Maize', 11–12.

[83] McCann, *Maize*, 27–29.

floury varieties were developed for breads, and softer varieties were developed to eat off the cob.[84]

Experimentation with maize was fairly short for a variety of reasons. It was attractive to locals in part due to the similarity of processing methods used for millet and sorghum grain heads, becoming an acceptable substitute to these older grains in all the same recipes, including porridge, fermented "breads", and beverages. Accordingly, techniques of the body, or a bodily logic that is socially transmitted, facilitated the incorporation of maize into the cuisine.[85] In addition to familiar processing methods, maize filled an ecological niche in the south of Ghana after the sixteenth century, especially in the forested centre.[86] When intercropped with cassava, maize helped convert the fragile forest soils into usable calories by protecting the forest floor from erosion. Requiring minimal tillage, maize was planted after the heavy March rains, allowed to tassel, and then intercropped with cassava and cocoyam, which preserved moisture for the maize, and protected the delicate forest floor from intense sun and hard rain.[87] Maize was early maturing, less laborious to tend, and provided two harvests, doubling the yield and leading to its domination of both forest and coastal eco-culinary zones.[88] Once it arrived from the Americas and was disseminated to the forest zone in the sixteenth century, maize and cassava together solved the problem of how to prevent vegetative regrowth in forests, keeping residual-clearing labour needs low. Maize and cassava offered the forest zone a continuous, year-round calorically dense food source. These additional low-labour calories enabled the feeding of major state systems, such as the Asante, who utilised maize as a transportable food for their army in the period of imperial expansion in the eighteenth and early nineteenth centuries.[89]

Maize also enabled local entrepreneurs and nobles on the coastal grasslands to profit from the Atlantic economy by capitalising on the demand for maize in forts and on ships. In order to meet this demand, local

[84] La Fleur, *Fusion Foodways*, 90–95.

[85] Marcel Mauss, 'Techniques of the Body', *Economy and Society* 2, no. 1 (February 1973): 75.

[86] Portères, 'L'Introduction du Maïs'.

[87] McCann, *Maize and Grace*, 44–49.

[88] McCann, *Green Land*, 121–22.

[89] McCann, *Maize and Grace*, 49.

leaders developed control over the arable land and required their subjects to tend to their plots, selling on the surplus to the Europeans. By the seventeenth century, local noblemen were using slave labour to develop plantations to supply townspeople, soldiers and craftspeople. They traded their maize for imported luxury foods and goods and redistributed these goods in harvest ceremonies that featured maize. In so doing, they created networks of dependency around the control and sale of maize to Europeans.[90]

The Portuguese in casting about to find other low labour, high yielding food sources for the provisioning of slaves, considered plantain as one of the earliest alternatives, but it was not an instant hit. This was due to the sometimes-unsuitable areas chosen to grow plantain on the coast. Also, the fragile nature of the young plants meant that marine transport from the Central African kingdom of Kongo (present-day Angola and Democratic Republic of the Congo) was particularly hard on them. The Kongo slaves brought by the Portuguese in the early sixteenth century were ultimately successful in plantain propagation, and plantain eventually proved attractive to locals in the woodlands north of Elmina, and areas like Axim and Shama, where the Portuguese also maintained forts. Plantains reproduced vegetally—a type of asexual reproduction where a new plant grows from a fragment of the parent plant—similar to the way yam reproduced. Another attractive feature for locals was that plantain required less labour to grow and to process and could be harvested year-round.[91]

New world varieties of cassava also came to be appreciated as a major part of the local diet. Initially, cassava grew in coastal areas exposed to sea spray and high winds where other foods, like maize, would not grow.[92] Toxins found in cassava needed to be processed out in order to render it edible. Also, once processed, cassava did not suit locals in terms of taste or texture. Aside from being laborious to process, cassava did not easily lend itself to all the customary ways starches were prepared, whether it was fermented, boiled, roasted, or pounded into a dense dumpling to go

[90] La Fleur, *Fusion Foodways*, 131–33.

[91] La Fleur, 55–73.

[92] Bosman, *Description,* 111–12.

with soup. For these reasons, cassava was considered the food of unfortunates. Seventeenth-century attempts to adapt it to develop more desirable qualities, like hardiness, and to reduce its high poison content were almost abandoned by the nineteenth century.[93] However, push factors such as drought, population pressure, disease and warfare caused the eventual adoption of cassava as a reliable bulky starch with which to fill empty bellies.

One by one, the coastal populations found themselves beginning to rely upon cassava due to its hardiness, versatility, and storability. Cassava could remain stored underground for long periods of time. The Fante, who endured a sequence of destructive Asante invasions between 1806 and 1816, were forced to turn to cassava and thereafter incorporated it into their gardens as insurance against hunger.[94] Declining soil fertility in the central forested eco-culinary zone also caused farmers outside the environs of the Asante capital Kumasi to turn to cassava. They discovered that it grew well on depleted soils and began to market it to growing communities of foreigners, slaves, and slave descendants in Asante.[95] By the early twentieth century, a combination of population pressure and disease caused cassava's adoption to become even more widespread. Cocoa farmers used migrants to work their farms and were looking for a crop to feed their workers that would not rob the cocoa of its nutrients. Additionally, the influenza pandemic of 1918–1919 depleted the ranks of farmers, necessitating the production of a less laborious crop.[96] Cassava was the crop local people increasingly employed to deal with these pressures.

---

[93] Reindorf, *History*, 267; Doku, *Cassava in Ghana*, 16.

[94] Hans Christian Monrad, *Two Views from Christiansborg Castle*, trans. Selena Axelrod Winsnes (Legon, Accra: African Books Collective, 2009), 53; Brodie Cruickshank, *Eighteen Years on the Gold Coast of Africa* (London: London, Hurst and Blackett, 1853), 128.

[95] Nana Akwasi Abayie Boaten, 'Asante: The Perception and the Utilization of the Environment before the Twentieth Century', *Institute of African Studies Research Review* 6, no. 2 (1 January 1990): 25; H. Bramer, 'Soils', in *Agriculture and Land Use in Ghana*, ed. J. B. Wills (London: Oxford University Press, 1962), 17.

[96] D. C. Ohadike, 'The Influenza Pandemic of 1918–19 and the Spread of Cassava Cultivation on the Lower Niger: A Study in Historical Linkages', *Journal of African History* 22, no. 3 (July 1981): 379–91; Doku, *Cassava in Ghana*, 19.

As with maize, the adoption of plantain proved to have wider economic and cultural consequences for local populations. Plantain, which required less labour to grow and process than yam, significantly freed up time for forest farmers to be redeployed toward gold mining. With plantain production taking on an increasingly important economic role in terms of feeding the miners, nobles and entrepreneurs became wary of the potential for plantain to become endowed with commensurate cultural value. As a counter to this potential, "big men" tightened control over yam production and reified yam culture, connecting its harvest to the spiritual landscape, effectively undercutting plantain culture. Not only was this control class based, but it was gendered in that a hierarchy was established whereby yam was cast as a male food that only men could propagate. Water yam and plantain were deemed as the food of women, children, or slaves.[97]

Coastal Region Staples

As previously stated, rainfall in the region around Accra was relatively low. The terrain constituted open grasslands in which tree crops, such as oil palms, fail to thrive.[98] These ecological circumstances encouraged trade with the interior and with Europeans, and resulted in the evolution of a cosmopolitan culture for people like the Gã. Historical food security issues on the coast were reflected in Gã cultural events such as the *Homowo* festival, where the urban population and the outlying villages drew together. *Homowo* (from *homo yi womo*—meaning to hoot at hunger) has dominated the Gã ritual calendar since the Gã migration to the coast, in which great privation was experienced.[99]

Key to the culinary introduction and acceptance of maize bread by the local populations, the Allada slaves brought by the Portuguese from the Slave Coast to the east were credited with the production of fermented

[97] La Fleur, *Fusion Foodways*, 108–11.

[98] Parker, *Making the Town*, 2–5.

[99] Field, 'Gold Coast Food', 10–11; Parker, *Making the Town*, 7.

maize bread called *kenkey*.[100] Both the Fante and Gã consumed *kenkey*, prepared as a fermented corn cake, and often wrapped in a banana leaf and baked in a small metal mobile oven. The Gã on the coast later adapted the Allada recipe by first wrapping the fermented maize in a maize husk instead of a banana leaf and then steaming the packet. The Fante prepared their kenkey by adding salt before folding their kenkey in a packet made of plantain leaves.[101] Other foods eaten by the Gã as noted by Danish Chaplain H. C. Monrad included the following: "Flatta, a porridge made of maize; palm nut soup made with fresh, smoked or salted fish or meat; kokobessa, chicken stewed in palm oil and the root of a plant; flamqvav, a kind of vegetable in palm oil, with dried fish, schattoe [shitto, a hot pepper sauce], and eggs; kasiokkel soup made with fish or meat; cabuseer kankies [kenkey]."[102] These same items are still eaten on the coast today, reflecting considerable continuity with regards to foodways.

The consumption of a thin maize meal porridge for breakfast on the coast is still extant today. In the early nineteenth century, maize meal porridge was sold near Christiansborg Fort in public places and towns in the early morning hours.[103] This porridge was considered a light breakfast. The heartier breakfast option for the Gã on the coast could include leftover kenkey or *ampesi* (boiled cassava) with leftover fish. The hearty main meal tended to be composed of protein rich stews with peppers. The menu did not vary from lunch to dinner, except for the after-dinner addition of roasted corn or groundnuts.[104]

Another essential staple was fish, critically important as a source of protein and as a flavouring for soups and stews. Fish was prepared either fried in palm oil, smoked, or salted. Rømer described a good meal as fish rich and meat poor. Palm oil—the key cooking oil on the coast—Spanish

---

[100] William Hutton, *A Voyage to Africa* (London: Longman, Hurst, Rees, Orme, and Brown, 1821), 101–02; La Fleur, *Fusion Foodways*, 121–23.

[101] La Fleur, 124.

[102] Monrad, *Two Views*, 184–85.

[103] Ibid., 184.

[104] M.J. Field, *Social Organization of the Gã People* (Accra: Crown Agents for the Colonies, 1940), 59.

piment, and kenkey with fried or dried fish was thought to be a satisfying meal.[105]

Goats and sheep were prevalent but considered expensive to eat. Animals were not milked, so virtually no milk-based foods were consumed on the coast. This may be due to the local view that milk products were considered unhealthful.[106] Hot peppers and palm oil, however, were thought beneficial to one's health. Rømer, a Danish merchant, was a keen observer of Gã culture on the coast in the 1740s. He observed mallaguet, piment (a type of pepper), palm oil, and citrons being used by female slaves aboard ship to prepare African medicines.[107]

Central Forested Region Staples

In the nineteenth century, the forested region was comprised of Akan-speaking kingdoms, dominated by the Asante. Blessed with a rich landscape that was home to the sacred Lake Bosomtwi, and comprised of lush jungle intersected by rivers, the region produced much fruit, fish, kola, yams, and gold. The delicate forest floor was susceptible to damage by intense sun and rain. A three-year crop rotation cycle was instituted, and different intensities of farm-use were adopted. During the first year, the most demanding crops, maize and yam, were planted on rich or replenished soil. During the second year, cocoyam and plantain were intercropped. Plantain trees helped protect the delicate soil from the intense sun. By year three, cassava was planted on the exhausted soil. Year four was the fallow year. This allowed the Asante a variety of foods, each of which carried a cultural weight, yam being the weightiest. This system was administered by matriclans who owned the land and used slave sharecroppers to perform the labour.[108]

As previously discussed, the cultivation of yams was reified by Akan male elders. As a result, by the nineteenth century *fufu* (a heavy, elastic

---

105 Rømer, Reliable Account, 196.

106 Monrad, *Two Views*, 127.

107 Rømer, Reliable Account, 199.

108 Ivor Wilks, *Forests of Gold: Essays on the Akan and the Kingdom of Asante* (Athens: Ohio University Press, 1995), 43–56.

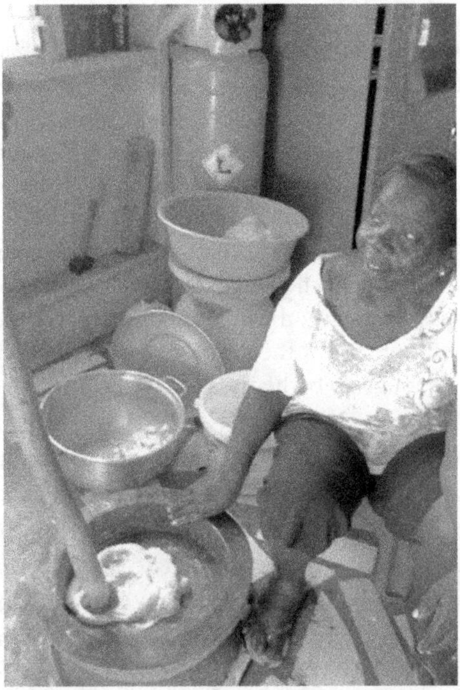

Fig. 2.1 Celina and partner pounding fufu at PRAAD Canteen, Accra

dumpling served with soup), made of a combination of yam and plantain, was widely consumed and ritually celebrated in the forested region in the *Odwira* festival, a celebration of the yam harvest in Asante and other Akan kingdoms.[109] Below, is an early twentieth century description of the preparation of yam and plantain *fufu* with a mortar and pestle by Edmund Perregaux:

> Yams or bananas [plantains] have had time to boil, a white foam comes out of the black boiling pot. With a clever movement, our negress takes off the lid, grabs the hot pot and turns it over to empty its contents into a bowl of wood or clay. She calls one or two companions or her children, throws the cut pieces in a mortar and pounds them with a pestle! Brushed by exercised hands, the pestles rise and fall in rhythm, crushing the boiled yam. When

[109] Bowdich, *Mission*, 319.

> the pieces are well crushed, one of the women sits next to the mortar and while her companion continues to vigorously pound in cadence, she turns and turns the dough, without fear of receiving a blow on the fingers; it happens, but very rarely! She adds a few drops of water from time to time and soon gets a consistent dough. When this dough is sufficiently worked, they form balls as big as the two fists and all these balls are deposited in a large clay bowl or also in several small bowls, placed on the ground, naturally! Now comes the distribution of soup and meat; an important and delicate operation if any. Each guest receives his or her foufous (that is to say one or two balls of dough); the husband usually receives two, so it is not difficult. But the soup! ... But the meat! ... Madame pours into each bowl a little soup, just enough to bathe the base of the foufou...[110]

Clearly, *fufu* pounding took skill and was very time consuming. Indeed, Perregaux's full account leads us to understand that the entire process from assembling the necessary ingredients, gathering the fuel, and drawing the water took a woman most of the day. In addition to all this preparation, it indicates that the processing was a two-person job (Fig. 2.1).

Another starch frequently combined with native yam into *fufu* was cocoyam (or taro), used as a hedge against other crop failures. Cocoyam production greatly increased from the 1840s as a new variety from the West Indies—*Xanthosoma mafaffa*—began to replace the African variety. The African variety, *Colocasia escualenta*, contained high levels of the stomach irritant calcium oxalate.[111] Perhaps its position as a garden plant used as a hedge against other crop failures diminished its prestige in Asante. By the late nineteenth century, cocoyam's popularity had grown as it became a staple people relied upon to feed their children.

> The colocase...forms one of the main elements of the diet. It belongs to the family Aroïdées, close neighbour of Orchids. A garden plant (the cala); its broad leaves of a beautiful green let escape a beautiful white flower shaped cornet...It is the favourite food of children, so much so that it is also called: the foster mother, "*oyeumma*", or even: What dries up the crying of the child, "*abofra gyae su*", or even: "*Kobi sin n'adamfo*", the friend of a small

[110] Perregaux, 'Chez les Achanti', 23–24.

[111] T. McCaskie, *State and Society in Pre-Colonial Asante* (Cambridge: Cambridge University Press, 1995), 170.

piece of fish! This funny name means this: Even if you have meat in fact a very small fish, like the kobi, if you have *amankanis* as a dish of resistance, that's enough![112]

Thus, in the mid-nineteenth century, the cocoyam became a staple food that did not require the lengthy removal of toxins that cassava did, or the investment in growth, weeding, watering, and mounding of earth to keep it moist, like yam. It grew easily in kitchen gardens and had edible leaves that provided additional nutrients to a diet that was starch-rich but poor in green vegetables. Cocoyam sped up cooking times and provided yet another easily processed and calorically dense food with which to support a growing population of immigrant labourers.

In the nineteenth century, food was not limited to what was locally produced in Kumasi, the capital city of the Asante and home to the royal court of the Asantehene (Asante King). Being at the crossroads and controlling the flow of trade between the coast and the savanna via the maintenance of roads, Kumasi was in a unique position to receive the best of imported goods. A thriving nineteenth-century northern interior trade between the central forested region and the Hausa and Mossi supplied sheep, goats, fowl, coarse cotton cloths, coarse blankets, *bommo* (cassava cakes), and shea-butter to the Asante.[113] Citrus, honey, and groundnuts were also readily available in the markets.[114]

The variety of available foods in Kumasi in the early nineteenth century was observed by envoy Willem Huydecoper, an officer of Dutch and Fante heritage. He himself brought tea as a gift to the Asantehene complaining that "I don't know who told these people that tea is the best cure for all illnesses. His majesty and his elders are always asking me for tea, and I have only brought a small quantity with me."[115] The Asante court had access to the best trade items from Europe as well as trade items from the north. Huydecoper reported the regular consumption of pork,

[112] Perregaux, 'Chez les Achanti', 32–33.

[113] Kwame Arhin, 'Aspects of the Asante Northern Trade in the Nineteenth Century', *Africa* 40, no. 4 (1970): 363–73.

[114] House of Commons Parliamentary Papers, 1849 XXXIV [C. 399], Missions to the King of Ashantee and Dohomey, 27 April 1849.

[115] Huydecoper, *Diary*, 16.

plantain, fowl, palm wine, sheep, banana, yams, chicken, cow and malaguetta peppers by the court. The Asante successfully raised chickens, sheep, pigs, goats, ducks and turkeys. The consumption of meat however, in addition to being structured by class, was predicated upon age and ritual:

> But it is not much more than the chiefs and the grown-ups who concur in the luxury of this meat, and even more so every day. Sheep are usually reserved for fetish sacrifices or to pay for the many lawsuits that every self-respecting Negro agrees from time to time. This is the reason why it is often very difficult for us to obtain meat; the natives who do not have too much for themselves do not want to sell us anything.[116]

This account reveals that the consumption of meat in Asante was reserved for men. Women and children generally consumed much less animal protein. Women did raise chickens and ducks for their eggs, but they often sold them or used them to compensate priests.[117]

Savanna Staples

Foods typical of the savanna zone were grain based, composed of both millet and sorghum. Wild foods and tree produce were valuable sources of non-cereal foods, especially during the dry (or hungry) season.[118] Many traditional domestic units in northern Ghana were composed of polygamous joint families of a man and his sons with their wives and unmarried daughters. Homes were constructed around a central granary, which played a key role in ritual and social composition and reflected the participation of the entire family in the growing of grains. Available root vegetables included yams, frafra potatoes, and sweet potatoes. The meal was composed of starch made of whichever grain was available, served with a soup composed of groundnuts or okra with vegetables, leaves/greens, and meat. Women grew the ingredients for soup which included

[116] Perregaux, 'Chez les Achanti', 20.

[117] Monrad, *Two Views*, 124.

[118] Destombes, 'Nutrition', 123.

okra, squash, tomato, garden eggs, leaves used in soups, and pepper. Condiments included oil from the shea nut and groundnut, purchased at the local market. Garden vegetables were typically dried by women and packed into pots and baskets for use in the lean season when, at the end of the annual farming cycle and just before the new harvest, there might be a shortage of cultivated food, and gathering would become necessary.[119]

Conclusion

The evidence from my research strongly suggests that staple foods in Ghana—and even whole cuisines—are attached to places. The significance of this finding is that the variances in staple foods between microclimates corresponded with variances in languages and distinct polities in the three different major regions. Additionally, this finding demonstrates the considerable complexity and nuance of the local foodways. Traditional foods were traded intra-regionally, and when long-term biological relationships were established with new staple foods, these too were traded intra-regionally and in the global marketplace.

In establishing the regionality of Ghanaian cuisine and its connections to the global economy, I have set the stage for thinking about food and for how we understand collective local identities and their associated social and cultural issues. Furthermore, reflecting upon the foodways complicates the traditional historical ways of thinking about West African involvement in the Atlantic World economy as providers of slaves. It recentres the narrative from the impact of the slave trade on Africa, to African agency in the shaping of rhythms of change as they integrated new foods, traded these foods, and gave these foods a unique meaning in each regional cultural context.

These findings are of significance to food studies scholars and to Atlantic World historians. This food studies approach to Ghanaian history brings a complexity to each region's relationship to the transatlantic trade heretofore unexamined in Atlantic world history. This same approach can be applied to other countries and regions in West Africa.

[119] Goody, *Cooking*, 51.

Being attentive to these processes would greatly expand our knowledge of West African cultural history, and further the process of centring African perspectives in the narrative of Atlantic world history.

References

House of Commons Parliamentary Papers, hosted at https://parlipapers.proquest.com/parlipapers:
Colonial Economic policy (Hansard, 17 December 1940)
PP 1849 XXXIV [C. 399] Missions to the King of Ashantee and Dohomey

Primary Sources

W. Bosman, *A New and Accurate Description of the Coast of Guinea* (Cass & Co., London, 1967)
T.E. Bowdich, *Mission from Cape Coast Castle to Ashantee* (J. Murray, London, 1819)
B. Cruickshank, *Eighteen Years on the Gold Coast of Africa* (London, Hurst and Blackett, London, 1853)
O. Dapper, *Naukeurige beschrijvinge der Afrikaensche gewesten van Egypten, Barbaryen, Libyen, Biledulgerid, Negroslant, Guinea, Ethiopiën, Abyssinie* (Jacob van Meurs, Amsterdam, 1668)
P. De Marees, *Description and Historical Account of the Gold Kingdom of Guinea (1602)*. Translated and edited by A.v. Dantzig, A. Jones (British Academy, Oxford, 1987)
A.B. Ellis, *The E'we-Speaking Peoples of the Slave Coast of West Africa, Their Religion, Manners, Customs, Laws, Languages, &c* (Chapman and Hall, London, 1890)
M.J. Field, *Social Organization of the Gă People* (Crown Agents for the Colonies, Accra, 1940)
M.J. Field, Gold coast food. *Petits Propos Culinaire*, no. 43, 7–19, (1993) [orig. *Gold Coast Food* (Achimoto College Press, Achimoto, 1931)]
W. Hutton, *A Voyage to Africa* (Longman, Hurst, Rees, Orme, and Brown, London, 1821)

W. Huydecoper, *Huydecoper's Diary: Journey from Elmina to Kumasi 28th April 1816–18th May 1817*. Translated by W. Graham (University of Ghana, Legon, 1962)

P.E. Isert, *Letters on West Africa and the Slave Trade*. Translated by. S.A. Winsnes (Sub-Saharan Publishers, Legon, Accra, 2007)

A. Jones (ed.), *German Sources for West African History, 1599–1669* (Steiner, Wiesbaden, 1983)

H. Meredith, *An Account of the Gold Coast of Africa: With a Brief History of the African Company* (Longman, Hurst, Rees, Orme, and Brown, 1812)

H.C. Monrad, *Two Views from Christiansborg Castle*. Translated by S.A. Winsnes (African Books Collective, Legon, 2009)

E. Perregaux, Chez Les Achanti. Société neuchâteloise de géographie **17**, 7–387 (1906)

J. Rask, *Two Views from Christiansborg Castle*. Translated by S.A. Winsnes (African Books Collective, Legon, 2009)

R.S. Rattray, *Ashanti* (Clarendon Press, Oxford, 1923)

R.S. Rattray, *The Tribes of the Ashanti Hinterland* (Clarendon Press, Oxford, 1932)

C.C. Reindorf, *History of the Gold Coast and Asante* (Missionsbuchhandlung, Basel, 1895)

The Republic of Ghana, General economic data—March 2004. FAO fishery country profile, http://www.fao.org/fi/oldsite/FCP/en/GHA/profile.htm. Accessed 20 October 2019

Secondary Sources

P.K. Acheampong, Rainfall anomaly along the coast of Ghana. Geografiska Annaler **64**(3/4), 199–211 (1982)

G.A. Acquah, *The Fantse of Ghana: A History* (Hull University, 1957)

F. Agbodeka (ed.), *Handbook of Eweland: The Ewes of Southeastern Ghana*, vol 1 (Woeli Publishing Services, Accra, 1997)

J.M. Allman, V.B. Tashjian, *'I Will Not Eat Stone': A Women's History of Colonial Asante* (James Currey, Portsmouth, 2000)

S.B. Alpern, What Africans Got for their slaves: a master list of European trade goods. Hist. Afr. **22**, 5–43 (1995)

A. Anyidoho, M.E. Kropp Dakubu, Ghana: indigenous languages, English, and an emerging national identity, in *Language and National Identity in Africa*, ed. by A. Simpson, (Oxford University Press, Oxford, 2008), pp. 141–157

K. Arhin, Aspects of the Asante northern trade in the nineteenth century. Afr. J. Int. Afr. Ins. **40**(4), 363–373 (1970)
B. Asinyo, C. Frimpong, E. Amankwah, Original article: the state of cotton production in Northern Ghana. Int. J. Fiber Text. Res. 7, p. 153 (2015)
J. Atta-Quayson (ed.), *Macmillan Atlas for Ghana* (Macmillan Education Ltd., London, 1995)
A.K. Awedoba, The peoples of Northern Ghana. National Commission on Culture Website—www.ghanaculture.gov.gh (2006), http://lagim.blogs.brynmawr.edu/files/2015/03/The-Peoples-of-Northern-Ghana.pdf
M.A. Awo, Marketing and market queens: a case of tomato farmers in the upper east region of Ghana. PhD diss. Rheinischen Friedrich-Wilhelms-Universität, Bonn, 2010
R.B. Bening, Land policy and administration in Northern Ghana 1898–1976. Trans. Hist. Soc. Ghana **16.2**(1), 227–266 (1995)
P. Bourdieu, *Distinction: A Social Critique of the Judgement of Taste* (Harvard University Press, Cambridge, 1984)
P. Bourdieu, *Language and Symbolic Power* (Polity Press, Cambridge, 1992)
H. Bramer, Soils, in *Agriculture and Land Use in Ghana*, ed. by J. B. Wills, (Oxford University Press, London, 1962)
J.G. Christaller, *A Dictionary of the Asante and Fante Language Called Tshi* (Basel, Evangelical Missionary Society, 1881)
J. Destombes, Nutrition and chronic deprivation in the West African Savanna: North Eastern Ghana, c. 1930–2000. PhD diss. London School of Economics and Political Science, 2001
E.V. Doku, *Cassava in Ghana* (Ghana Universities Press, Accra, 1969)
M. Douglas, *Purity and Danger: An Analysis of Concepts of Pollution and Taboo* (Routledge, London, 1966)
M. Douglas, *Natural Symbols: Explorations in Cosmology* (Penguin, Harmondsworth, 1973)
P. Falk, *The Consuming Body* (SAGE, Newbury Park, 1994)
M.J. Field, The agricultural system of the Manya-Krobo of the gold coast. Africa **14**(2), 54–65 (1943)
C. Fischler, Food, self and identity. Soc. Sci. Inf. **27**(2), 275–292 (1 June 1988)
M. Fortes, S.L. Fortes, Food in the domestic economy of the Tallensi. Africa **9**(2), 237–276 (1936)
J.K. Fynn, *Asante and Its Neighbours, 1700–1807* (Northwestern University Press, Evanston, 1971)
R.S. Gocking, *The History of Ghana* (Greenwood, Westport, CN, 2005)

J. Goody, *Technology, Tradition, and the State in Africa* (Oxford University Press, London, 1971)

J. Goody, *Cooking, Cuisine and Class: A Study in Comparative Sociology* (Cambridge University Press, Cambridge, 1982)

S.E. Greene, *Gender, Ethnicity, and Social Change on the Upper Slave Coast* (Heinemann, Portsmouth, NH, 1996)

M.D.W. Jeffreys, Pre-Columbian Maize in Africa. Nature **172**(4386), 965–966 (November 1953)

R.A. Kea, Trade, state formation and warfare on the gold coast, 1600–1826. PhD. diss. University of London, London, 1974

M.E. Kropp Dakubu, K.C. Ford, The central Togo languages, in *The Languages of Ghana*, (International African Institute, New York, 1988)

J.D. La Fleur, *Fusion Foodways of Africa's Gold Coast in the Atlantic Era* (Brill, Leiden, 2012)

M. Lynn, The profitability of the early nineteenth-century palm oil trade. Afr. Econ. Hist. **20**, 77–97 (1992)

D.J.E. Maier, Precolonial palm oil production and gender division of labor in nineteenth-century gold coast and Togoland. Afr. Econ. Hist. **37**, 1–32 (2009)

M. Manoukian, *Akan and Ga-Adangme Peoples of the Gold Coast* (Published for the International African Institute by Oxford University Press, London, 1950)

M. Mauss, Techniques of the body. Econ. Soc. **2**(1), 75 (February 1973)

J. McCann, *Green Land, Brown Land, Black Land: An Environmental History of Africa, 1800–1990* (Heinemann, Portsmouth, 1999)

J. McCann, *Maize and Grace: Africa's Encounter with a New World Crop, 1500–2000* (Harvard University Press, Cambridge, 2007)

M. McCarthy, *Social Change and the Growth of British Power in the Gold Coast: The Fante States, 1807–1874* (University Press of America, Lanham, 1983)

T.C. McCaskie, *State and Society in Pre-Colonial Asante* (Cambridge University Press, Cambridge, 1995)

T.C. McCaskie, Denkyira in the making of Asante c. 1660–1720. J. Afr. Hist. **48**(1), 1–25 (2007)

E.L.R. Meyerowitz, *Akan Traditions of Origin* (Faber and Faber, London, 1952)

T. Naden, The Gur languages, in *The languages of Ghana*, ed. by K. Dakubu, (International African Institute, London, 1988), pp. 12–49

P. Nugent, "A few lesser peoples": The central togo minorities and their Ewe neighbours, in *Ethnicity in Ghana: The Limits of Invention*, ed. by C. Lentz, P. Nugent, (Palgrave USA, New York, 2000)

P. Nugent, A regional melting pot: the Ewe and their neighbours in Ghana-Togo, in *The Ewe of Togo and Benin*, ed. by B. N. Lawrance, (Woeli Publishing Services, Accra, 2005)

K. Ofosu-Budu, D.B. Sarpong, Oil palm industry growth in Africa: a value chain and smallholders' study for Ghana, in *Rebuilding West Africa's Food Potential*, ed. by A. Elbehri, vol. 44, (FAO/IFAD, Rome, n.d.)

D.C. Ohadike, The influenza pandemic of 1918–19 and the spread of cassava cultivation on the lower Niger: a study in historical linkages. J. Afr. Hist. **22**(3), 379–391 (July 1981)

G. Owusu, S. Agyei-Mensah, A comparative study of ethnic residential segregation in Ghana's two largest cities, Accra and Kumasi. Popul. Environ. **32**(4), 332–352 (1 June 2011)

J. Parker, *Making the Town: Ga State and Society in Early Colonial Accra* (James Currey, Oxford, 2000)

R. Portères, L'Introduction du Maïs en Afrique. Journal d'agriculture traditionnelle **2**(5), 221–231 (1955)

T.M. Reese, "Eating" luxury: Fante Middlemen, British goods, and changing dependencies on the gold coast, 1750–1821. William Mary Q. **66**(4), 851–872 (2009)

R. Tremblay, M. Landry-Cuerrier, M. Humphries, Culture and the social-ecology of local food use by indigenous communities in Northern North America. Ecol. Soc. **25**(2), p. 153 (April 2020)

M. Venkatachalam, *Slavery, Memory and Religion in Southeastern Ghana, c.1850–Present* (Cambridge University Press, 2015)

I. Wilks, *Forests of Gold: Essays on the Akan and the Kingdom of Asante* (Ohio University Press, Athens, 1995)

F. Willett, The introduction of Maize into West Africa: an assessment of recent evidence. Afr. J. Int. Afr. Inst. **32**(1), 1–13 (1962)

C.G. Wise, Climatic anomalies on the Accra plain. *Geography* **29**(2), 35–38 (1944)

Worldometer, Ghana population. Elaboration of data by United Nations (2020), https://www.worldometers.info/world-population/ghana-population/#:~:text=Ghana%202020%20population%20is%20estimated,of%20the%20total%20world%20population. Accessed 1 July 2020

J. Zimmermann, *A Grammatical Sketch of the Akra or Ga Language* (Basel Mission, Stuttgart, 1858)

3

The Proper Meal

"You have not eaten until you have eaten *fufu*." I heard this statement in Accra while chatting with the managing director of Y Semereka Limited. Ethnically Akan, he claimed he ate *fufu* each evening, and he would eat it three times each day if he could get it. This trained accountant from the Brong Ahafo Region responded to the question of what the national dish of Ghana is with "Depends on your tribe," adding that the Akan are in the majority, and they take *fufu*.[1] Those who eat fufu consider it their most characteristic and only completely satisfying food.[2] The Akan are not the only ethnic group to consider their local food the most satisfying and best food to eat:

> There are people who will say they haven't eaten the whole day, simply because they haven't had their soup and fufu. If you give them anything– bread sandwich, Caesar salad, they don't consider it as food, until they've

[1] Simpson Miller, 'Food and Nationalism', 77.

[2] Gracia Clark, 'Money, Sex and Cooking: Manipulation of the Paid/Unpaid Boundary by Asante Market Women', in *The Social Economy of Consumption*, eds. Henry J. Rutz and Benjamin S. Orlove (Lanham, MD: University Press of America, 1989), 332.

B. Simpson Miller, *Food and Identity in Nineteenth and Twentieth Century Ghana*, Food and Identity in a Globalising World,
https://doi.org/10.1007/978-3-030-88403-1_3

> sat down with their bowl of fufu and soup. Or fish and kenkey for the Gas (the Gã people).[3]

For those from the north, the staple would be *tuo zaafe* (or 'TZ'; stiff millet porridge) with soup, and the Ewe, in the Volta Region, would dine on *banku* (fermented, steamed maize dumpling).[4] The cultural weight of the local traditional food, its power to satisfy, and the overall collective social emphasis on the main meal of the day have persisted in Ghana, irrespective of centuries of trade with Europeans.

James McCann discerned that regardless of the ecological region, most cooking in West Africa was based upon starches that imparted a feeling of fullness: "Starches are…bland. Yet they provide an essential factor of texture, shape, and bulk that frames other components of a meal and is in fact the defining feature of the culinary culture."[5] As in other West African countries, Ghanaian cuisine was founded upon a starchy staple, and this foundation is precisely why Ghanaian cuisine has had such staying power. This chapter will discuss the history of the culture of cooking and eating in Ghana and what Ghanaians deemed to be a proper meal. The associated social systems and cultural values as expressed by the preparation and consumption of that meal will be examined. Last, the ways in which food was conceived of in stories and festivals will be reviewed to undergird the understanding of Ghanaian social organization and the persistence of Ghanaian foodways.

In Ghana, as in the west, a cooked meal at the end of the workday was considered a "proper" meal, necessary for the propagation of the health and welfare of the household.[6] The main meal, typically served in the late afternoon or evening, was a hot, cooked, one course, two or three-part meal, of which a heavy processed starch was the foundation. A two-part meal consisted of a soup and a dense starch, whereas a three-part meal

[3] Fran Osseo-Asare, Interview with Dinah Ayensu, 2018, https://betumi.com/library/http-betumi-com-library-articles-beyond-gumbo-a-history-of-ghanaian-cookbooks-by-fran-osseo-asare/interview-with-dinah-ayensu/.

[4] Helena Margaret Tuomainen, 'Migration and Foodways: Continuity and Change among Ghanaians in London' (PhD. diss., University of Warwick, 2006), 126.

[5] McCann, *Stirring the Pot*, 33.

[6] Anne Murcott, 'The Cultural Significance of Food and Eating', *Proceedings of the Nutrition Society* 41, no. 2 (1982): 207–08.

consisted of a dense starch served with a centre piece of fish and a sauce. M. J. Field constructed a matrix of the different starches in her 'General Survey of Gold Coast Food', that reflected the variety of processes applied to starches in the urban areas of the Gold Coast in the early twentieth century.[7] This survey stands out for its clear delineation between fermented starchy foods and unfermented starchy foods and their myriad preparation methods that included roasting, soaking, sprouting, fermenting, pounding, boiling, mashing, and mixing. Field made it clear that the preparation of the staple starch was often accomplished in multiple phases, at times taking hours or days to complete, and often requiring assistance.

Once processed, the starch was characteristically covered or accompanied by a hot soup or sauce which served as a companagium. A companagium is defined as follows: "What is meant here in particular is the 'ragout' type of dish, whose spicy and abundant sauce is scooped up with bread broken into pieces and held in the fingers; sometimes, too, the pieces of a pliable bread…help to catch and hold in the fingertips the pieces of food which the various guests take from the common dish."[8] This dense starch could be served warm or room temperature, but the soup should be hot. The soup should include meat flavouring, ideally combined with some fish, which can be fresh or smoked. The meat could be fresh or dried, fish or fowl, bush meat or domesticated. Oil was another critical ingredient. An oily sheen added substance to the meal and imparted health and strength.[9] The most desirable qualities of proper food were the ability to satiate hunger and the presence of a pleasing smooth and heavy texture of both the soup and starch, as proper food was usually swallowed, not chewed.[10]

---

[7] Field, 'Gold Coast Food', 11–18.

[8] Margaret L. Arnott, *Gastronomy: The Anthropology of Food and Food Habits* (Berlin: De Gruyter Mouton, 2011), 312.

[9] Perregaux, 'Chez les Achanti', 42; Christaller, *Dictionary*, 391.

[10] See: Fran Osseo-Asare, '"We Eat First With Our Eyes": On Ghanaian Cuisine', *Gastronomica* 2, no. 1 (February 2002): 49–57 and La Fleur, *Fusion Foodways*, 123. In fact, La Fleur observed that Gã-speakers invented a new name, k ɔ mi, (from k ɔ, 'bite' [habitual] + mi , 'swallow' [habitual]) which marked their appreciation of this bread and tells us that swallowing was how it may have been consumed.

The soup or sauce, like the starch, was also prepared in multiphasic stages. Again, Field's survey highlighted the processes involved in cooking the soup, including boiling, removal and mashing, stone grinding, pounding, and frying. Sometimes the meat was boiled first, especially if it was fresh. Usually, the vegetables were boiled for a time, then fished out and ground, and then reintroduced to the soup and cooked for a time more. The soup should have a satisfying thickness that was achieved with the addition of ground dried seeds, groundnuts, or dried shrimp, depending upon what was locally available. Pepper, tomato, oil, and onion have become essential ingredients for the soup, supplemented by local vegetables such as garden egg (local aubergine), or the leaves from local greens.[11] The whole preparation of the meal could take up to six hours, depending on the distance between home and market, as the shopping trip could take two to three hours, the same time spent cooking.[12]

The emphasis of a proper meal was placed on the cooking, and the techniques employed required the cook to closely monitor the proceedings. There was nothing convenient about a proper meal. Fuel and water needed to be fetched at the outset. Depending upon the starch used, it could be boiled in progressive stages until very stiff and might require constant stirring (millet, sorghum). Alternatively, the starch may have needed grinding, fermenting and steaming (maize). Or the starch might have needed boiling and pounding until the desired smooth texture was reached (yam or plantain for example), which often necessitated the labour of more than one person and required strength and coordination. All the techniques employed required labour, planning and time, right up until the moment of service, to ensure the proper temperature.[13] It was not a requirement that the meal be commensal.[14] There were many circumstances where a family would not eat together due to work or school obligations, or ritual observances or restrictions. What was

[11] Dinah Ayensu, *The Art of West African Cooking* (Garden City: Doubleday, 1972); Dede, *Ghanaian Favourite Dishes*.

[12] Clark, 'Money', 333.

[13] Jack Goody, *Cooking, Cuisine and Class: A Study in Comparative Sociology* (Cambridge: Cambridge University Press, 1982), 69.

[14] Helena Tuomainen, 'Eating Alone or Together? Commensality among Ghanaians in London', *Anthropology of Food*, no. S10 (26 November 2014): 3.

important was that the food be provided by the woman, who oversaw its preparation. Food work and the ability to cook an acceptable meal were considered foundational to becoming an adult woman. Women's agency was visible in processes of domestic food allocation, in the provisioning and/or withholding of food.[15] Women could express anger or dissatisfaction with a marriage by serving small portions, spoiling the food by oversalting the soup, not serving the food in a timely fashion, or refusing to cook altogether.[16] There was an agency to how a woman served food.[17] The ability to control which people in her sphere received food and the amount they received, was an expression of a woman's priorities and a way to reward behaviour she saw as beneficial (Fig. 3.1).

Traditionally, the meal was eaten with the hands, so handwashing before eating was customary. Today, in restaurants that serve local food, you will find a handwashing station complete with a sink, hand (or dish) soap, and towels. Food was eaten only with the right hand as the left was considered unclean. The starch was pinched off between index, middle finger, and thumb. An indentation was made in the starch with the thumb, which was then dipped in the soup or sauce and then swallowed, not chewed. The meat was often picked out and chewed separately.[18]

As a rule, one ate with people of similar social status according to age, sex, rank, and social role. The phrase: 'you are what you eat' has a necessary corollary: 'you are with whom you eat.'[19] Based on the Latin *com-* "together with" + *panis-* "bread" a companion was literally "one who breaks bread with another".[20] The transgression of these boundaries was

---

[15] Abbots, *Agency*, 50–52; Jon Holtzman, 'Politics and Gastropolitics: Gender and the Power of Food in Two African Pastoralist Societies', *Journal of the Royal Anthropological Institute* 8, no. 2 (2002): 259–78; Williams-Forson, 'Other Women', 448.

[16] Clark, 'Money', 327; Clark, 'Fasting', 48–49.

[17] Cornelia Nell, 'Commensality and Sharing in an Andean Community in Bolivia', in *Commensality: From Everyday Food to Feast*, eds. Susanne Kerner, Cynthia Chou, and Morten Warmind (London: Bloomsbury, 2015), 172.

[18] Osseo-Asare, 'We Eat First'; Osseo-Asare, *Food Culture*, 33; Fortes and Fortes, 'Food'; Williams-Forson, 'Other Women', 5; Goody, *Cooking*, 77.

[19] Jordan Rosenblum, 'Justifications for Foodways and the Study of Commensality', in *Commensality: From Everyday Food to Feast*, eds. Susanne Kerner, Cynthia Chou, and Morten Warmind (London: Bloomsbury, 2015), 193.

[20] 'Companion, n.1', in *OED Online* (Oxford University Press), accessed 12 May 2020, https://www.oed.com/view/Entry/37402.

Fig. 3.1 Handwashing station at Buka Restaurant, Osu, Accra

considered especially humiliating. In a 1923 Akropong Native Affairs inquiry into the treatment of a captain who had committed a wrongdoing, a witness observed the following punishment; "This captain was forced by the accused to eat with small children and the captain has to clean the dish."[21] This punishment of a senior man of rank being made to eat with little children was considered "contrary to native manners", and clearly illustrated the strict social rules governing eaters in the Gold Coast.

In Ghana, it was customary for men to be served first and the mothers and children later. Men ate alone or with other men from a common dish. Often, boys would eat in the company of the men, although in a separate group which shared the same bowl. This situation presented an opportunity for a senior male to express approval or affection of a junior by rewarding the junior with an extra portion of meat. Women ate with other women and/or with children. In some areas, there existed no cultural requirement that husband and wife reside together in the same household or even pool their finances. Many resided with the matrilineal clans and patrilineages with whom they were affiliated. Often, food was

[21] 'Aburi Native Affairs Destoolment Inquiry' (Akropong, 1923), ADM KD 29/6/21, PRAAD Accra.

delivered to the husband's house by a child from the place where the wife resided with her family.[22]

Food flows up in Ghana. What is meant by this is that the size of portions was structured by one's age and gender.[23] Older men customarily got the best of the food prepared. This could be the woman's brother, father, or husband, depending upon her obligations and stage of life. It was primarily for the benefit of the man that proper food was prepared.[24] For a woman's social inferiors, any food do. Even an absence of protein was not considered to be remiss in the meals of the children. "An older word for 'mother' [in Ewe] is *da* or *dada*, which appears to be derived from the verb *da*, 'to cook'; it being the mother who prepared the food for the family. The terms for 'mother,' then, like those for 'father,' had no relation to the act of begetting or of birth; but meant 'she who stays in the house,' or 'she who cooks.'"[25] Only proper food would be accepted by a husband, and a proper meal was cooked by a woman.[26]

The transformation of starch into heavy food was woman's work (at times aided by her social inferiors), while the man's job was to supply her with chop money—the funds to prepare the cooked dinner.[27] In contemporary parlance, the woman would be called the "chop master"—the one who purchases the food and sets the menu.[28] If men did cook, it was usually in a ritual capacity and then the food was roasted, not boiled, and nothing was ever ground by a man. Traditionally, men were the butchers, cutting up and cleaning the meat before it was handed over to women for cooking. An exception to this rule was the cleaning and preserving of fish, which was usually done by women.[29]

---

[22] Field, *Social Organization*, 60; Meyer Fortes, *Time and Social Structure and Other Essays* (London: University of London Athlone Press, 1970), 5; Clark, 'Money', 327; Stephan F. Miescher, *Making Men in Ghana* (Bloomington: Indiana University Press, 2005), 30.

[23] Goody, *Cooking*, 68.

[24] Tuomainen, 'Migration', 130.

[25] Ellis, *Ewe-Speaking Peoples*, 214–15.

[26] M. J. Field, *Search for Security: An Ethno-Psychiatric Study of Rural Ghana* (London: Faber and Faber, 1960), 28; Miescher, *Making Men*, 242 n131.

[27] Blench and Dendo, 'Dictionary', 12; Christaller, *Dictionary*, 76.

[28] Blench and Dendo, 'Dictionary', 12.

[29] See: Koley Ambah and Ga Manste Tackie Obiri, 'Fish Ovens—Destruction of by Accra Town Council', February 1909, ADM 11/1/78, PRAAD Accra; Field, *Social Organization*, 62.

The husband did not just provide chop money, but the exchange of chop money for meals created other social bonds and obligations. The cooking of food by a woman for a man signified a sexual relationship.[30] The nineteenth-century Twi word *di* or *dzidzi* meant both to eat or to have sexual intercourse with. Today the word "chop" is used in a similar fashion both as a noun (meaning food) and as a verb—to have sex with.[31] For fufu eaters, the pestle represents the male part when pounding *fufu* is used as a metaphor of heterosexual intercourse.[32] There were broader socio-economic obligations that came with this relationship. Women became obligated to assist the husband in the marketing of his business. If he was a fisherman, she would help him to sell and/or process his catch. The husband likewise had obligations to his wife to assist with expenses for their children, such as the payment of school fees, or help with clearing the farm before planting.[33] The biggest obligation borne by a woman was the preparation of proper food for a husband, a testament to her having spent her time in a wifely fashion. The nature of proper food—its mode of preparation, the demand that the woman spend time procuring the ingredients, the care with which those ingredients were prepared, and the requirement that the meal be served in a timely manner—epitomised the wife's obligation to the husband.[34]

Food intersected not just with gender and seniority, but also with ethnicity. The proper meal revealed much about one's regional identity and the social role one occupied in relation to that identity. The word "can" was frequently used in relation to food, as in, "Can you eat that type of food?"[35] Your ability to eat a type of food revealed your regional affiliation, family group, clan, lineage, and your social status. For example, an Akan might not feel able to eat cassava depending upon the way it was prepared. Many identified themselves positively as consuming certain

---

[30] Clark, 'Fasting', 51; Jean Marie Allman and Victoria B. Tashjian, *'I Will Not Eat Stone': A Women's History of Colonial Asante* (Portsmouth: James Currey, 2000), 60.

[31] Blench and Dendo, 'Dictionary', 12.

[32] Clark, 'Fasting', 51.

[33] Field, *Social Organization*, 64; Fortes and Fortes, 'Food', 9.

[34] Clark, 'Money', 326.

[35] Blench and Dendo, 'Dictionary', 11; Miescher, *Making Men*, 79.

foods, needing it prepared in a specific way due to their ethnic affiliation.[36] What one could eat also located one upon the social scale in terms of age and seniority. For example, a senior high-status male might frown upon eating fried plantain, claiming this food was not sufficiently "heavy" enough for him.[37] Thus the ideal meal in Ghana was complicated by a system of classifications, rules of propriety, and contexts that ordered the world and gave it meaning.[38]

Just as there was an ideal meal, there were ideas about what an ideal body looked like, and about the role that proper food played in helping to construct that body. These ideas have persisted over time to the present day. While travelling in Accra with two corpulent middle-aged men (one of whom was my husband), I overheard them conversing with each other about strangers' reactions to their bodies. Our companion remarked that he felt like a rock star as random strangers would approach him to praise his physique. Once a middle-aged Ghanaian man upon seeing him exclaimed, "I like your body!" My husband too had received several approving comments about his size and prodigious belly from random men on the street. Big bodies represented wealth, plenty and the ability to feed others. Heavy, starchy food was appreciated for its ability to create big bodies and to fortify the form.[39] The word fufu literally means white in Twi.[40] For many Akan and Gã the whiteness of the starch symbolised victory and joy.[41] A big body represented victory over hunger and the ability to support many dependents.

---

[36] Tuomainen, 'Migration', 126; T. E. Kyei, *Our Days Dwindle: Memories of My Childhood Days in Asante*, ed. Jean Allman (Portsmouth: Heinemann Educational Books, 2001), 126.

[37] Clark, 'Fasting', 49; Huydecoper, *Diary*, 74.

[38] Claude Fischler, 'Food, Self and Identity', *Social Science Information* 27, no. 2 (1 June 1988): 285, https://doi.org/10.1177/053901888027002005.

[39] Osseo-Asare, 'We Eat First', 53.

[40] Christaller, *Dictionary*, 138.

[41] G. P. Hagan, 'A Note on Akan Colour Symbolism', *Research Review* 7, no. 1 (1970): 8–14; Regina Kwakye-Opong, 'Clothing and Colour Symbolisms in the Homowo Festival: A Means to Sociocultural Development.', *Research on Humanities and Social Sciences* 4, no. 13 (2014): 112–25.

Myths, Legends, and Festivals

Thus far, the proper meal has been discussed, the social systems and cultural values that underlie it have been considered, and the commensurate pressures to prepare the proper meal and the opportunities to supply others with sustenance have been explored. In this section we will examine some myths, legends, and festivals to see how they structured ideas about cooking and eating, and how these practices fed beliefs, reconstituted the body, and reaffirmed social identities. The construction and recitation of myth and legend are reflected in one's practices, values, attitudes, and ideologies, and are literally embodied in the food that is eaten.[42] Pierre Bourdieu's concept of *habitus* allows us to see the body as an object of historical inquiry that can be approached through every day practices such as cooking and eating.[43] Using the work of Rebecca Earle and William Rubel as inspiration, I will attempt to relay the food as story—the symbolic value-creation and social construction of memory based upon myth, legend, and festival—as it is embodied in the daily practices of cooking and eating on the Gold Coast.[44]

Food is always enmeshed in narrative: "All foods, as opposed to raw ingredients, are products of culture and thus imbedded in the realm of words; whenever we eat, we eat a story."[45] The following is a retelling of the significance of the Sacred Grove at Asantemanso in the central forested region:

This grove outside Kumasi is the most hallowed place in Ashanti [Asante] territory. At this spot, Ashanti myth affirms, the first human beings, belonging to certain of their ruling clans, came forth from the ground and settling nearby, increased and multiplied, learned to use fire and other arts, until eventually, compelled by increasing numbers, they

[42] Rebecca Earle, *The Body of the Conquistador* (Cambridge: Cambridge University Press, 2012), 2–3.

[43] Pierre Bourdieu, *Distinction: A Social Critique of the Judgement of Taste* (Cambridge, Mass.: Harvard University Press, 1984), 170; E. M. Collingham, *Imperial Bodies: The Physical Experience of the Raj, c.1800–1947* (Cambridge: Polity, 2001), 3.

[44] Mintz and Du Bois, 'The Anthropology of Food and Eating', 99–119; William Rubel, 'Food as Story: Story as Food', in *Food and Language: Proceedings of the Oxford Symposium on Food and Cooking 2009*, ed. Richard Hosking (Devon: Prospect Books, 2010); Collingham, *Imperial Bodies*; Earle, *The Body of the Conquistador*.

[45] Rubel, 'Food as Story: Story as Food', 293.

scattered and became clan or "blood" from which the rulers of the united nation later chose their kings and queens.[46]

Legend has it that a worm bored its way up to the surface at Asantemanso and led seven men, several women, a leopard, and a dog out of the earth. The story goes that the people were afraid and their eyes roamed about wildly in fear. One of the men, Adu Ogyinae, laid hands upon them to soothe them. A couple of days later, while busying themselves constructing huts, Adu Ogyinae was killed by a tree they were felling. He became an honoured ancestor. After this, the dog left and came back with fire in its mouth. Food was laid upon the fire, and the dog was fed with the food to ascertain whether it was fit for consumption. The dog grew fat upon this food, and as it grew fat, men came to eat cooked food. The first ancestors then settled at Nampansa, where the soil is red. The creator then took one of their own, a linguist, with him to the sky. From that time long ago to this, the King of Asante sent a cow each year to be sacrificed at that spot where the ancestors came from the ground.[47]

In this origin story, food was one of the main ways that the supreme being communicated with his human creations, and they with each other. This story of place tells us that this grove was initially dangerous and frightening. The death of one of their companions reminded them of their relationship to the place, and their weakness in relation to their creator. Much like Adam and Eve in the Garden of Eden, this new and strange grove had the effect of making them conscious of their place in the social order, and their dependency on God. Their relationship to each other and to their animal companions (via the assistance of their dog helper) was made clear by the phrase, "As it grew fat, the men came to eat cooked food". This part of the myth helps us to understand that humans differ from animals in their cooking of food. It also suggested that it was the men who came to eat the cooked food first, and by omission, implied the women were doing the cooking. Well along in the myth, the linguist was taken by the creator. Ritual sacrifices were then typically made to intercessory ancestors like the linguist to facilitate communication with the creator. The origin myth of Asantemanso clearly delineates the right

[46] R.S. Rattray, *Ashanti* (Oxford: Clarendon Press, 1923), 121–24.

[47] Rattray, *Ashanti*, 121–24.

relationships between the men and the women, between them and the animals, their creator, and their ancestors, all through food. It is still customary for Akan women to observe the practice of serving men first and of ensuring his needs are anticipated: "The women see to it that the men's food is properly set before [the men] are invited to come [in]. Not that [the men] sit [start eating] and later ask for water, for a piece of salt and so."[48] These relationships then are recreated and reinforced with each main meal, every sacrifice, and each harvest festival.

In a different ecological region, just east of present-day Accra, a proper meal established and maintained the identity, safety, and continuity of the Sukpe community. Colonial district record books from 1912 documented the oral history of the Dangme-speaking peoples' migration from Nigeria to the mouth of the Volta River on the Atlantic.[49] At the time, oral histories were collected as part of the Crown Lands Bill of 1897, meant to ascertain how tribes came to acquire the lands they currently inhabited. Following documentation of the oral histories, boundaries were mapped, and certificates of validity were issued to facilitate negotiations for gold concessions. The following migration story reflected the need for local people to maintain their connection to and ownership of the land, with regular sacrifices of food.

A cooking contest won the migrating Sukpe control over the vital salt and fishing trade on the Volta River, a trade which would sustain them throughout the cessation of the transatlantic slave trade in the nineteenth century. On an urgent survey expedition, the *wulomei* (priest) and hunter Lomo-weir encountered the Ewe-speaking Ahaviatse, another expeditionary hunter from nearby Agavedon, on the eastern side of the Volta. They disputed fervently among themselves as to who first saw the land and creeks thereon, including the nearby Nyito creek. Both resignedly decided that each would go to his home to bring back fire. No sooner had Ahaviatse left than Lomo-weir furtively took flint from his hunting bag, made a fire, and cooked some fish from the creek. Lomo-weir may have fried this fish in palm oil with Spanish peppers, or perhaps cooked a

[48] Miescher, *Making Men*, 29.

[49] G.A.A. Okansey, 'The Discovery of the Okor Forest and the Songor Lagoon', District Record Book, ADM 44-5-1 (PRAAD, Accra, 1912).

ragout of fish with hibiscus and palm oil. He ate and left some for Ahaviatse. Upon the Ewe hunter's return, he found chop ready for him. The dispute was resolutely settled in favour of Lomo-weir. The victorious Sukpe gained the lands from the Mi valley in the west, inclusive of the land between the lagoon and the coast, all the way to the Volta in the east. The hunters willingly befriended each other, with Ahaviatse choosing to resettle his people near Lomo-weir's, creating the settlement named Big Ada.[50] Big Ada would become especially prosperous in the nineteenth century due to its trade in palm oil, fish, and salt. To this day, fried tilapia and *banku* (steamed fermented maize dumpling) are known as proper Ewe food.

How did Lomo-weir manage to win the land with its vital fishing rights and salt production capabilities simply by being the first to cook supper? The Sukpe wielded ritual food preparation as a weapon with which to gain advantage over the Ewe-speaking inhabitants near the Volta River. As a priest, Lomo-weir's function was to read the will of local gods and spirits, and appeal to them for assistance. A priest was also charged with the establishment and maintenance of shrines for this purpose. Food was the most common sacrifice at these shrines. Lomo-weir would have observed ritual while hunting and preparing food, in sacrifice and thanks to local deities. In a place and time where sin was equated with the neglect of deities, iniquity itself could be expunged, and land ownership was established, by the offering of food sacrifice. Thus, proper food was contingent upon observances of the moral and ethical dimensions of the local spiritual landscape.

In the case of the Sukpe, Lomo-weir's ritual food observances helped to further ensure the survival and stability of his people in their flight to safety. This historical and ritual cooking contest empowered the Sukpe to rightfully establish a stronghold in a place far away from the slave raids of the ever-expanding Akan empire to the west. Later in the twentieth century, the story of this cooking contest was accepted by the British colonial government as a foundation for the demarcation of Sukpe lands. Proper

[50] Brandi Simpson Miller, 'The Moral and Ethical Aspects of Gold Coast Foodways', *Gastronomica: The Journal of Critical Food Studies* 19, no. 1 (Spring 2019): 111–12; District Record Book 1912, ADM 44/5/1, PRAAD Accra.

food to the Sukpe was ritual food that had powerful spiritual significance, and which was used to give them the influence and authority to make a new place for themselves on the coast. In this migration story, food, and its transformation into nourishment with fire, decided claims about the division of resources and determined, in part, relations among people. Food's ritual use as a spatial marker was testimony to the power of the transformative properties of cooking. As a reminder of the crucial role cooking plays in the story of being human, this history shows how proper food and cooking enabled eighteenth-century West Africans to lay claim to, and to create a future for themselves.

In the neighbouring Ga region, long standing food security issues on the coast, especially in the months of June and July, were reflected in cultural events such as the *Homowo* festival. A celebration of the harvest of the first maize of the season, *Homowo* was the time of year when *kpoi-kpoi* (unfermented, steamed maize meal seasoned with a palm oil and okra sauce) and soup were served as a ritual food.[51] A celebration of victory over hunger in many rural communities, *Homowo* began with the planting of crops before the rainy season near the coast from May to August. Bans on noise and drumming allowed the gods to concentrate on producing an abundant harvest. The ritual meal day, called *koyeligby*, happened sometime in August or September.

The coast was composed of frontiers of consumption that separated the coastal towns with their salt lagoons and fish, from the hinterland with its offerings of maize and palm oil.[52] The Gã landscape was organised into two spheres: one domain of fishing people on the coast, and another of farming hamlets seven miles or so inland, where the soil was more fertile. Six coastal towns ruled a hinterland that produced maize and palm oil, which were traded for fish and salt with their friends by the sea. The fishing villages were regarded as senior to those in the hinterland. This was due to the seniority of the coastal *Kpesi* god *Sakumo*, who was honoured above the gods of immigrants from the hinterland. Maize figured prominently in *kpoikpoi* (or *Sakumo's kpekpei)*, the ceremonial food

---

[51] Carl Christian Reindorf, *History of the Gold Coast and Asante* (Basel: Missionsbuchhandlung, 1895), 277.

[52] Jones, *Feast*, 190–222.

which, on *Homowo* Tuesday was cooked in Sakumo's honour at his shrine by male elders and spiritual leaders.[53]

The *Homowo* feast reinforced and replicated social relationships, affirmed identities and promoted integration, harmony, allegiance, and reconciliation.[54] The *Nai wulomo* (chief priest) was guardian of the community's spiritual health and head of the Gã ritual hierarchy.[55] For *Homowo*, each farming village community brought its produce to the coastal towns and in exchange could expect to receive support and protection from the towns in times of need. These communities were internally held together by redistribution and externally by warfare and trade. This was the essence of the *Homowo* redistributive feast, which bound the community together.[56]

In addition to reinforcing political and trade affiliations, the feast replicated the spiritual order. On the feast day, the *Nai wulomo* sprinkled *kpoikpoi* and asked for intercession from the gods of the land and sea for the Gã people. The *mantsemi* (literally the 'father of the town', singular *mantse*), who were secular officeholders, then sprinkled food around the community to purify it and to invite the ancestors and deities to dine with the people. After these ceremonies, families would gather round the family hearth to eat with the ancestors.

The soup, prepared by an appointed group of women, brought together the four main ingredients of coastal and hinterland consumption (palm oil, salt, fish, maize), whose combination into a ritual dish symbolised unity, harmony, trade, and cooperation. Its red colour derived from palm oil and symbolised the qualities they wanted to uphold as a community. In Gã cosmology, red represented danger and aggression, but also denoted pride, love, caution, bravery, determination, hard work, and alertness. White, derived from the kenkey, symbolised purity, virtue, joy, and victory—in this case over hunger.[57] Eating the *kpokpoi* was meant to promote

[53] Field, *Social Organization*, 142–64.

[54] Dominique Juhé-Beaulaton, 'L'alimentation Des Hommes, Des Vodoun et Des Ancêtres' in *Cuisine et société en Afrique: histoire, saveurs, savoir-faire*, ed. Monique Chastanet, François-Xavier Fauvelle-Aymar, and Dominique Juhé-Beaulaton (Paris: Editions Karthala, 2002), 53.

[55] Field, *Social Organization*, 82, 143.

[56] Jones, *Feast*, 190–91.

[57] Kwakye-Opong, 'Clothing and Colour', 114–20.

increased opportunities for success and blessings for a new year. The meal also served to consolidate the Gã people around the extended family hearth, located at the heart of their cultural landscape and ritual topography, represented by the coastal towns, such as Accra, the sea and its god Nai, and the lagoons and their gods.[58] The sharing of *kpokpoi* at the *Homowo* feast reaffirmed a sense of place, knitting the Gã together as a people that would stand collectively against outsiders.

In a society of historical ethnic mixing, sharing ritual foods played a key role in the process of becoming Gã. In performing the important rituals, the Nai wulomo reaffirmed his authority as a spiritual figurehead who substantiated the relationship between the dead, the gods, and the living. Further, these communal meals were novel because they transformed the diners, their relationship with the environment, and their relationship to each other.[59] Held together by redistribution and externally by warfare and trade, Gã unity was celebrated and reaffirmed in the lavish feast of *Homowo*.

The following version of an origin story of the Kuulo people of Banda on the forest-savanna frontier recalls the tradition of migration that is also shared by the Ewe, Gã, and Akan peoples further south. It encapsulates the centrality of food and cooking to the integration of strangers and the peaceful cohabitation with ethnically divergent neighbours in the Northern Territories. The Nafana hunter Gbaha arrived from the north-east of present-day Côte d'Ivoire from the town of Kakala, where the people were migrating due to a chieftaincy dispute. Saying that he was an emissary from Kakala, he begged fire from Wurache. She obliged, but three days later Gbaha returned to Wurache's settlement while everyone was away, taking fire and extinguishing her hearth. When Wurache returned she wept for she had no fire on which to cook, whereupon God sent a messenger to ask why she wept. The messenger provided a chain that carried her to the sky, and she was given fire and returned to earth. Gbaha returned to Wurache's settlement to fetch fire again. Eventually, cordial relations prevailed between Wurache and Gbaha's people, and

[58] John Parker, *Making the Town: Ga State and Society in Early Colonial Accra* (Oxford: James Currey, 2000), 21, 170.

[59] Jones, *Feast*, 215–16.

Wurache gave her daughter, Akosua Yeli, to Gbaha in marriage. As the population of Nafanas grew, Wurache and Gbaha met and agreed that the Nafana leader would be chief, though Wurache retained ownership of the land.[60] Ownership of the land signified that the gods of Wurache's people were given primacy over the gods of newcomers.

The story about the ancestress Wurache being the keeper of the hearth, and therefore of local heritage, also reflected a certain flexibility and fluidity with regards to identity and social and economic composition. The influential place of food in social structure was demonstrated in the following commentary by Brillat-Savarin: "The table established a kind of alliance between the parties and made guests more apt to receive certain impressions and submit to certain influences. This was the origin of political gastronomy."[61] The story of the hearth fire and Wurache's negotiation with Gbaha over who would retain spiritual governance over the land's produce was a narrative found among many ethnically divergent neighbours all over Ghana. The forging of a peaceful alliance of indigenous farmers and migrating outsiders using food as a means of cohesion, is reflected in rituals performed in northern Ghana to this day.

Conclusion

My research makes the historical cuisines of Ghana evident by illuminating the structure of the proper meal. The rules of a proper meal in Ghana dictated how, when, and by whom specific foods were prepared, who got to eat which foods, and with whom one ate. Only the main meal founded upon a starchy staple could provide the desired feeling of satiety because only proper meal provided physical, spiritual, and cultural nourishment to a household. Starchy staples were sacred substances that imparted spiritual strength, reconstituted the blood, and fortified the body. The concept of terroir has been and remains very real in Ghana. Just as Spanishness

[60] Ann Brower Stahl, *Making History in Banda: Anthropological Visions of Africa's Past* (Cambridge University Press, 2001), 148–88.

[61] Jean Anthelme Brillat-Savarin, *The Physiology of Taste: Or Meditations on Transcendental Gastronomy: Transcendental Gastronomy* (Scotts Valley: CreateSpace Independent Publishing Platform, 2017), 25.

was a bodily construct that defined the difference between Spanish conquistadors and the inhabitants of the Americas, being Gã, Ewe, or Asante was also a bodily construction that required proper food to identify the cultural borders between them.[62]

This consideration of the proper meal and its attendant meanings, and the examination of the oral history and how it revealed the ways food was used to structure right relationships, expands and brings depth and interest to the social history of Ghana. This chapter addresses a gap in the scholarship whereby West Africa was relatively unexplored in terms of how food facilitated African understandings of personal and collective identity and the understanding of relationship to social Others. The application of interdisciplinary food studies methods to social history in Ghana enables us to think about a host of historical and contemporary social and cultural issues such as the role of transnational trade in the conception of self-identity. Ghana is reimagined as a place where distinct ways of being were mediated by microclimates and based on staple foods. This approach and methodology serve to centre the West African gaze in Atlantic world history.

References

E.-J. Abbots, *The Agency of Eating: Mediation, Food and the Body* (Bloomsbury Publishing, London, 2017)

Aburi native affairs destoolment inquiry. Akropong, ADM KD 29/6/21. PRAAD Accra, 1923

J.M. Allman, V.B. Tashjian, *'I Will Not Eat Stone': A Women's History of Colonial Asante* (James Currey, Portsmouth, 2000)

K. Ambah, G.M.T. Obiri, Fish ovens—destruction of by Accra town council, ADM 11/1/78. PRAAD Accra, February 1909

M.L. Arnott, *Gastronomy: The Anthropology of Food and Food Habits* (Walter de Gruyter, Netherlands, 2011)

D.A. Ayensu, *The Art of West African Cooking* (Doubleday, Garden City, 1972)

[62] Earle, *The Body of the Conquistador*, 53–65, 146–50.

R. Blench, M. Dendo, "Observers are worried": a dictionary of Ghanaian English. Ghana English Dictionary (2006), http://www.rogerblench.info/Language/English/Ghana%20English%20dictionary.pdf

P. Bourdieu, *Distinction: A Social Critique of the Judgement of Taste* (Harvard University Press, Cambridge, MA, 1984)

J.A. Brillat-Savarin, *The Physiology of Taste: Or Meditations on Transcendental Gastronomy: Transcendental Gastronomy* (CreateSpace Independent Publishing Platform, Scotts Valley, CA, 2017)

J.G. Christaller, *A Dictionary of the Asante and Fante Language Called Tshi* (Basel, Evangelical Missionary Society, 1881a)

J.G. Christaller, *A Dictionary of the Asante and Fante Language Called Tshi* (Evangelical Missionary Society, Basel, 1881b)

G. Clark, Money, sex and cooking: manipulation of the paid/unpaid boundary by Asante market women, in *The Social Economy of Consumption*, ed. by H. J. Rutz, B. S. Orlove, (University Press of America, Lanham, MD, 1989)

G. Clark, *Onions Are My Husband: Survival and Accumulation by West African Market Women* (University of Chicago Press, Chicago, 1994)

G. Clark, From fasting to fast food in Kumasi, Ghana, in *Food Consumption in Global Perspective: Essays in the Anthropology of Food in Honour of Jack Goody*, ed. by J. Klein, A. Murcott, (Palgrave Macmillan, Basingstoke, 2014)

E.M. Collingham, *Imperial Bodies: The Physical Experience of the Raj, c.1800–1947* (Polity, Cambridge, 2001)

Companion, n.1. In *OED Online*. Oxford University Press, https://www.oed.com/view/Entry/37402. Accessed 12 May 2020

A. Dede, *Ghanaian Favourite Dishes* (Anowuo Educational Publications, Accra, 1969)

District record book, ADM 44/5/1. PRAAD Accra, 1912

R. Earle, *The Body of the Conquistador* (Cambridge University Press, Cambridge, 2012)

A.B. Ellis, *The Eʻwe-Speaking Peoples of the Slave Coast of West Africa, Their Religion, Manners, Customs, Laws, Languages, &c* (Chapman and Hall, Limited, London, 1890), http://archive.org/details/cu31924028624926

M.J. Field, *Social Organization of the Gã People* (Crown Agents for the Colonies, Accra, 1940)

M.J. Field, *Search for Security: An Ethno-Psychiatric Study of Rural Ghana* (Faber, 1960)

M.J. Field, Gold coast food. Petits Propos Culinaire (43) 7–19 (1994)

C. Fischler, Food, self and identity. Soc. Sci. Inf. **27**(2), 275–292 (1 June 1988). https://doi.org/10.1177/053901888027002005

M. Fortes, *Time and Social Structure and Other Essays* (University of London, Athlone Press, 1970)

M. Fortes, S.L. Fortes, Food in the domestic economy of the Tallensi. Africa **9**(2), 237–276 (1936)

J. Goody, *Cooking, Cuisine and Class: A Study in Comparative Sociology* (Cambridge University Press, Cambridge, 1982)

G.P. Hagan, A note on Akan colour symbolism. Res. Rev. **7**(1), 8–14 (1970)

J. Holtzman, Politics and gastropolitics: gender and the power of food in two African Pastoralist societies. J. R. Anthropol. Inst. **8**(2), 259–278 (2002)

W. Huydecoper, *Huydecoper's Diary: Journey from Elmina to Kumasi 28th April 1816–18th May 1817*. Translated by W. Graham (Legon: University of Ghana, 1962)

M. Jones, *Feast: Why Humans Share Food* (Oxford University Press, Oxford, 2008)

D. Juhé-Beaulaton, L'alimentation Des Hommes, Des Vodoun et Des Ancêtres, in *Cuisine et Société En Afrique, Histoire, Saveurs, Savoir-Faire*, ed. by M. Chastanet, F.-X. Fauvelle-Aymar, D. Juhé-Beaulaton, (Editions Karthala, Paris, 2002), pp. 53–66

R. Kwakye-Opong, Clothing and colour symbolisms in the Homowo festival: a means to sociocultural development. Res. Humanit. Soc. Sci. **4**(13), 112–125 (2014)

T.E. Kyei, *Our Days Dwindle: Memories of My Childhood Days in Asante*. Edited by Jean Allman (Portsmouth: Heinemann Educational Books, 2001)

J.D. La Fleur, *Fusion Foodways of Africa's Gold Coast in the Atlantic Era* (Brill, Leiden, 2012)

J.C. McCann, *Stirring the Pot: A History of African Cuisine* (Ohio University Press, Athens, OH, 2009)

S.F. Miescher, *Making Men in Ghana* (Indiana University Press, Bloomington, 2005)

B. Miller, Food and nationalism in an independent Ghana. MA Thesis, Georgia State University 2015

S.W. Mintz, C.M. Du Bois, The anthropology of food and eating. Annu. Rev. Anthropol. **31**, 99–119 (2002)

A. Murcott, The cultural significance of food and eating. Proc. Nutr. Soc. **41**(2), 203–210 (1982)

C. Nell, Commensality and sharing in an Andean community in Bolivia, in *Commensality: From Everyday Food to Feast*, ed. by S. Kerner, C. Chou, M. Warmind, (Bloomsbury, London, 2015)

G.A.A. Okansey, The discovery of the Okor forest and the Songor Lagoon. District Record Book. PRAAD, Accra, 1912

F. Osseo-Asare, "We eat first with our eyes": On Ghanaian cuisine. Gastronomica **2**(1), 49–57 (2002)

F. Osseo-Asare, *Food Culture in Sub-Saharan Africa* (Greenwood Press, Westport, 2005)

F. Osseo-Asare, Interview with Dinah Ayensu, 2018., https://betumi.com/library/http-betumi-com-library-articles-beyond-gumbo-a-history-of-ghanaian-cookbooks-by-fran-osseo-asare/interview-with-dinah-ayensu/

J. Parker, *Making the Town: Ga State and Society in Early Colonial Accra* (James Currey, Oxford, 2000)

E. Perregaux, Chez Les Achanti. Société neuchâteloise de géographie **17**, 7–387 (1906)

R.S. Rattray, *Ashanti* (Clarendon Press, Oxford, 1923)

C.C. Reindorf, *History of the Gold Coast and Asante* (Missionsbuchhandlung, Basel, 1895)

J. Rosenblum, Justifications for foodways and the study of commensality, in *Commensality: From Everyday Food to Feast*, ed. by S. Kerner, C. Chou, M. Warmind, (Bloomsbury, London, 2015)

W. Rubel, Food as story: story as food, in *Food and Language: Proceedings of the Oxford Symposium on Food and Cooking 2009*, ed. by R. Hosking, (Prospect Books, Devon, 2010)

B. Simpson Miller, The moral and ethical aspects of gold coast foodways. Gastronomica: J. Crit. Food Stud. **19**(1), 111–112 (Spring 2019)

A.B. Stahl, *Making History in Banda: Anthropological Visions of Africa's Past* (Cambridge University Press, 2001)

H.M. Tuomainen, Migration and foodways: continuity and change among Ghanaians in London. PhD diss., Warwick, 2006

H. Tuomainen, Eating alone or together? Commensality among Ghanaians in London. Anthropol. Food (S10) (2014). https://doi.org/10.4000/aof.7718

P. Williams-Forson, Other women cooked for my husband: negotiating gender, food, and identities in an African American/Ghanaian household. Fem. Stud. **36**(2), 435–461 (2010)

4

The Asante and Diplomatic Use of Food: A Symphony of Signals

Between May 1816 and March 1820, the King of Asante, or Asantehene, Osei Tutu Kwame (a.k.a. Osei Bonsu; r.1800–1824), received no less than nine diplomatic missions at Kumasi, the royal capital in the forested centre of Ghana.[1] This chapter examines the written accounts of feeding and feasting during these audiences with the Asantehene. It maps the interface between Europeans and their Asante hosts in the context of commensality and the exchange of food from 1816 to 1848. The diplomatic use of food and dining was mobilised as a political tool in trade negotiations, and food was a ubiquitous apparatus in the art of Asante statecraft in the nineteenth century. The Asante use of food was intended to influence politics and trade, to define relationships of superiority and inferiority, or to communicate messages of symbolic kinship.[2]

The reason for the diplomatic attention paid to the Asante by the Europeans on the coast was connected to the recurrent Asante military invasions of the coastal Fante region between 1806 and 1816. These

[1] Fiona Sheales, 'Sights/Sites of Spectacle: Anglo/Asante Appropriations, Diplomacy and Displays of Power 1816–1820' (PhD diss., University of East Anglia, 2011), 3.

[2] Linda Morgan, 'Diplomatic Gastronomy: Style and Power at the Table', *Food and Foodways* 20, no. 2 (April 2012): 146.

B. Simpson Miller, *Food and Identity in Nineteenth and Twentieth Century Ghana*, Food and Identity in a Globalising World,
https://doi.org/10.1007/978-3-030-88403-1_4

conflicts were largely a result of the contraction of the slave trade (the British abolition of the slave trade was in 1807) and the establishment of what was termed the legitimate trade for palm oil. The Asante invaded the coast to control trade with the Europeans more directly, causing major disruptions to business. Europeans had a vested interest in stabilising their position on the coast and in developing new markets for manufactured goods beyond the forested centre. Therefore, direct diplomatic contact with Asante power in Kumasi was necessary.

Commensality is the embodied experience of social relations with people. What follows is an exploration of the diverse nature of signs and symbols associated with nineteenth-century Asante diplomatic gastronomy. The first section outlines the social structure and forest foodscape of Asante in the nineteenth century. The next section examines three state dinners and several diplomatic residencies hosted by two Asantehenes in Kumasi. Drawn from richly detailed primary sources, these encounters show how the Asante weaponized food as a diplomatic tool through preparation, table setting, seating and gifting. The chapter concludes with an examination of the historical implications of changes in courtly food culture between the time of Osei Tutu Kwame and Kwaku Dua Panin (1834–1867) and will consider how the introduction of new foods transformed cooking in this region in the nineteenth century.

Asante Foodways

Yam played an important role in Asante cosmology and a central role in the tale of how Onyame (God) left this world and withdrew to live in the sky. Onyame was said to have been happy to live among humans but for the actions of one greedy old woman. This old woman, having invented the mortar and pestle, disturbed Onyame with her incessant pounding of fufu. Each time she lifted and dropped her pestle, she pushed him further and further away into the sky. Realising what she had done, she entreated her granddaughter and other villagers to help her stack the mortars into a ladder to reach Onyame. Eventually, they realised they did not have enough mortars so the old woman suggested that they remove the bottom mortars and stack them on top, thinking they could reach Onyame this way. Unfortunately, the mortar ladder toppled and killed many people. The people went away aggrieved and disappointed, and they never attempted to reach Onyame this way again.[3]

A noteworthy aspect of this story is its gendered implications. Women, engaged in their archetypal occupation of cooking, were the cause of the retreat of Onyame. Traditionally, women fetched wood and water, farmed with their family's assistance, and prepared food.[4] Women's sin of greed and their lack of wisdom while performing their duties resulted in the need for the intercession of the ancestors with Onyame. Moreover, women's capriciousness rendered them unfit to communicate with ancestors and is the reason that men became the primary conduit through which effective communication with ancestors was made. This was reflected in the social and ritual hierarchy of Kumasi that was headed by the Asantehene and managed by his largely male ancillary staff.

The role of cooks in the Asante royal household was reflective of this concern to manage the relationship between humans and Onyame via the ancestors. The imperial cooks—*soodoofoo*—assisted the Asantehene in ensuring that the three simultaneous planes of existence—the dead, the living, and unborn—were all able to eat and prosper.[5] Ritual interplay between humans and deities involving sustenance was managed by the Asantehene and his *soodohene* (head cook). The department of cooks was fundamental to the ritual purity and wellbeing of the Asantehene and was a part of the *nhenkwaa* (servants of the King) that numbered in the thousands and had daily domestic and public roles and functions. The *nhenkwaa* evolved over time into the instrument through which the Asantehene's will found expression with the outside world and facilitated the Asantehene's contact with the ritual world. Serving under the *soodohene* were subgroups that include *boodeedwafo* (cutters of plantain), *fufuwofo* (pounders of fufu), *nfohoofo* (roasters of meat), *mansufo* (providers of drinking water), and *akwanmofo* (chewing stick providers). Over time,

---

[3] R. S. Rattray, *Ashanti Proverbs: The Primitive Ethics of a Savage People* (Oxford: Clarendon Press, 1916), 146–47.

[4] Claire C. Robertson, *Sharing the Same Bowl?: A Socioeconomic History of Women and Class in Accra, Ghana* (Bloomington: Indiana University Press, 1984), 56–67; De Marées, *Description*, 22.

[5] Joseph K. Adjaye, *Diplomacy and Diplomats in Nineteenth Century Asante* (Lanham: University Press of America, 1984), 20.

the *soodoofoo* became an integral part of the standardised Asante diplomatic process and held a prominent place in the procession of the king's retainers.

Evidence from the nineteenth century revealed that Kumasi was shaped by a gendered hierarchy that was reflected in ritual using food. Special commensality, such as that encountered during the Odwira festival, served to further establish power relations that were central to the maintenance of political organisation and community, providing necessary communal labour.[6] In a performative act of gendered fealty, at the Odwira festival vassals pledged their loyalty to the Asantehene with the means with which to produce food (slaves) and the result of food production (produce).[7] The fealty oath of "marriage" between Asantehene and vassal was formalised by the presentation of a grinding stone, signifying that Kumasi and the Asantehene were to "eat" from the vassal. The Dutch officer Willem Huydecoper was witness to this relationship between the King and his vassals. A Fante second lieutenant, merchant, diplomat and grandson of one-time Director-General Jan Pieter Theodoor Huydecoper, Huydecoper wrote on 19 November 1816 that at midnight, the King summoned him to discuss the rebellion of the Wassaws. General Tando had not asked the Wassaw for the terms of their surrender, and the Asantehene wanted the Dutch government at Elmina to make a treaty with the Wassaws on the Asatehene's behalf. The King explained the problems with Tando's treaty thus:

> That it was the custom among the negroes that, when a woman commits adultery and injures her husband thereby, separates from him for a time and then wishes for a reconciliation, she first sends a sheep, chickens, etc., on before her, and then later comes herself to settle the real matter at issue; Tando in his negotiations with the Wassaws has not stipulated any of these things and has thus not been able to collect any damages from the Wassaws.[8]

[6] Susanne Kerner and Cynthia Chou, 'Introduction', in *Commensality: From Everyday Food to Feast*, ed. Susanne Kerner, Cynthia Chou, and Morten Warmind (London: Bloomsbury, 2015), 6.

[7] Ivor Wilks, *Asante in the Nineteenth Century: The Structure and Evolution of a Political Order* (London: Cambridge University Press, 1975), 295.

[8] Huydecoper, *Diary*, 56.

In this account, the Asantehene communicated his understanding of his relationship between himself and his wayward vassal as one of a broken marriage. Food in this account served as the foundation for knowledge of the world and its right order. When casting the Wassaw in the role of adulterous wife, this reflected both the generally held Asante understanding of the obligation of women to provide food and the place of the Wassaws in the social hierarchy as subordinate to the Asante.

Fufu

Plantain, yam, or cocoyam, boiling water, salt, onions, pepper

- Roughly peel the starch of your choice, cut into small pieces, and place them in a large black terrine. Add a little water, salt, pepper, some onions, and cover, allowing the starch to boil
- Boil from an hour to an hour and a half, and then begin the accompanying soup composed of vegetables with some chicken, sheep, pig, or goat
- Once the yams or bananas have had time to boil, a white foam will come out of the black boiling pot. Remove the lid, grab the hot pot, and turn it over to empty its contents into a bowl of wood or earth.
- Place the cut pieces of starch in a mortar and use the pestle in rhythmic movement to crush the boiled yam. This is best performed with a companion who turns and turns the dough. Periodically add a few drops of water from time to time to produce a consistent dough.
- When this dough is sufficiently worked, form balls as big as two fists deposit in a large bowl or in several small bowls
- Pour a little soup into each bowl, just enough to bathe the base of the fufu, being careful to ensure the husband receives the lion's share. Repeat until all the soup has been distributed. Repeat the same procedure with the bits of meat in the soup.

Adapted from Perregaux, 'Chez les Achanti', 22–24.

The principal staple food for the Asante was the yam. It would not be an exaggeration to say that the rise of the Akan forest kingdoms was a direct result of the symbiotic relationship between yam and man. The African yam zone runs along a north-south area from the Atlantic coast to the frontiers of the savanna, and from an east-west axis, beginning roughly at present-day Congo to present-day Guinea. The introduction of ironworking into West Africa roughly 2,500 years ago facilitated deeper penetration into the forest, where improved forms of wild yams were increasingly favoured over grain crops like millet. Millenia of propagation reduced the fertility of yam, *Dioscorea*, rendering it dependent upon human promulgation, much like maize in the Americas became dependent upon human intervention. Yam cultivation provided the Akan with a more adequate, reliable, and generally superior nutritional basis in the forest zone, allowing for the elaboration of philosophical notions that ennobled the tuberous cultivar.[9] Yam cultivation was the province of men and was governed by older men at the top of the Asante social hierarchy who organised the workers. Requiring significantly more toil than plantain, yam farming mandated additional social control over the workforce. "Big men"—or social elites—who organised yam production were invested in the celebration of yam culture as part of the ritual landscape and committed resources to celebrate the yam harvest and their special relationship to its cosmic significance.[10]

The very name of the yam dish fufu is a derivative of the Akan word for white, *fufuo*, which symbolised the colour of purity and sacredness. A ritually auspicious colour, *fufuo* is also associated with victory and spiritual purity. It is the colour of gods and of kings, symbolising joy, hope and wellbeing. A person's *kra*, the metaphysical aspect of a person associated with one's destiny, is associated with the colour of fufuo.[11] Yam, and the food produced from it, is considered a prestigious food and carries with it the associations of its colour. Consuming fufu was both physically and spiritually satisfying.

As outlined in Chap. 1, in the seventeenth and eighteenth centuries, the adoption of plantain facilitated the transformation of Asante to an ever-expanding agricultural empire capable of supporting a large influx of labourers from the savanna region.[12] The association of wealth with labour with which to grow a surplus was still extant in the nineteenth century, even after the Atlantic slave trade was in decline. Notwithstanding

[9] D.G. Coursey, 'The Origins and Domestication of Yams in Africa', in *West African Culture Dynamics: Archaeological and Historical Perspectives*, ed. B. K. Swartz and Raymond E. Dumett (Berlin: De Gruyter, Inc., 1980), 85.

[10] J. D. La Fleur, *Fusion Foodways of Africa's Gold Coast in the Atlantic Era* (Leiden: Brill, 2012), 111.

[11] Hagan, 'Akan Colour Symbolism', 8–14.

[12] T. C. McCaskie, *State and Society in Pre-Colonial Asante* (Cambridge: Cambridge University Press, 1995), 38.

plantain's importance as the main starchy staple, yam was the foundational forest crop whose regulation, production, and ritualised consumption underwrote Asante power. McCaskie's definitive analysis of the *Odwira* Festival (commonly known as the Yam Custom in the nineteenth century) argued that the cultivation of yams was mobilised to cement and extend the power of the state in its largesse and its ability to protect its subjects from hunger and harm. The Odwira (from *dwira*: to cleanse, to purify) was a harvest festival whose performance assured the continued prosperity of the empire via the ritualised concentration and expulsion of defilement. Defined by McCaskie as "a festival of cleansing and purification, orientated towards a ritual mediation on the seamless unity of dead, living, and unborn," the ritualised eating of new-year yams at the Odwira was a highly structured ceremony whose timing was determined by the cycle of rains and the harvest.[13] Such festivals were the major socio-religious event of the year throughout the West African yam zone, held to mark the time when the new crop could first be eaten. The ritual prohibition period where yams were not allowed to be eaten corresponded with the phase of the active growth of the yam plant during the rainy season, during which it was highly susceptible to damage. It is around the protection of this food resource that other sanctions and rituals began to develop.[14]

Formalised under the first Asantehene Osei Tutu, at the time of the inception of the Asante Kingdom in the late seventeenth and early eighteenth centuries, the annual Odwira reinforced state control over food and was the point of departure for conditions of material and social reality, ordering space and time and establishing the right order of history for the Asante. Ritualised interruptions of daily life before the Odwira with farming bans and the imposition of bans on the eating of yams, enforced reflection on the place of Asante culture in nature and individual identity within Asante.[15] Therefore, the sacramental consumption of yams was used to create a culture, a set of standardised values of which the Asantehene was the executor.

[13] McCaskie, *State and Society*, 146.

[14] Coursey, 'The Origins and Domestication of Yams', 83.

[15] McCaskie, *State and Society*, 153.

The Asantehene's social function was to administer and to perform ritualised strategies that addressed the symbolic struggle for sustenance. It was a performative magic designed to elicit supernatural protection against the possible infertility of the earth. Bowdich observed the Odwira of 1817 and recounted that "All the chiefs kill several slaves that their blood may flow into the hole from whence the new yam is taken."[16] This ritual was to ensure the future abundance of yam harvests. The Asantehene's absolute control of the timing of the first yam harvest (no one would harvest until the Asantehene gave his approval, upon pain of death), and the performance of the consumption of newly harvested yams, portended the total sacramental regulation of food and its supply by the state.[17]

The ritual observance that closed the Odwira, the *odwira nwonakwasie*, was performed at a critical juncture of the agricultural calendar. It was the perfect example of the liturgical effort at the creation of social magical conditions that represented a "reciprocal transaction, in which the products of human cultivation and therefore culture were offered back to nature in order to…acknowledge their mutual interdependence."[18] The *odwira nwonakwasie* was one such reciprocal transaction performed under cover of night whereby the new yams, smeared on one side with a red substance and the other with black, were processed by the Asantehene and his party to a crossroads where the Cape Coast road crossed the wet ground near the Nsuben River. A linguist addressed the entity known as *awo* (the first, hermaphrodite sacrifice to *asase yaa*—the earth—to ensure fruitfulness) and all the spirits of deceased foes (*asamanfo*). "He announced the celebration of the Odwira, asked the *asamanfo* to reconcile themselves to their assigned status, requested them to come and eat of the offering of new yam, and asked that all future enemies might suffer the same fate as themselves in the interests of the preservation and progress of Asante."[19] The new yams were then thrown into the bush toward the *asamanfoe* and *awo*, followed by a volley of shot and the silent and swift return of the

[16] Thomas Edward Bowdich, *Mission from Cape Coast Castle to Ashantee* (London: J. Murray, 1819), 228.

[17] McCaskie, *State and Society*, 147.

[18] Sheales, 'Sights/Sites of Spectacle', 126.

[19] McCaskie, *State and Society*, 222.

procession home. The addressing of *awo* as the first sacrifice to ensure the fruitfulness of the earth is absolutely revealing of the principal purpose of the Odwira. This ritual officiated by the Asantehene designated him as the maintainer of relations between the realms of the living, dead, and the unborn.

Fufu played a key role as the dish desired above all others, and the desire for fufu was structured and reflected in the social organisation. Huydecoper observed and recorded the variety of available foods in Kumasi and complained bitterly of his frequent consumption of plantain: "I am reduced to eating fried plantains and there is great distress in my house."[20] Yam, plantain, rice, and cocoyam (taro) were the starchy foundations of meals in Kumasi. However, plantain and cocoyam were considered food for slaves: "The lady takes out her provisions; she withdraws either bananas [plantain], yams, or colocases [taro or cocoyam], according to her taste or according to the importance of her guests; the yam is for the husband or for the guests of choice, the banana for her, the children or the slaves."[21] Plantain could be eaten boiled, fried, or, combined with yams, to make a satisfying fufu that was served with soup. Irrespective of its ability to satisfy, there were clear ideas about the low rank plantain occupied in the hierarchy of foods in nineteenth-century Kumasi. Huydecoper reflected these attitudes in his commentary about the distress of eating food intended for women and slaves.

Much like the adoption of plantain by the Akan in the seventeenth century, the adoption of maize was a boon as well, and was credited with the rapid expansion of the Asante empire north into the southern savanna in the eighteenth-century.[22] A transportable forest-based food supply for the military, ground maize was easy and light to carry and could quickly be prepared into a nourishing porridge or a portable loaf with just a bit of water and salt. Captured Basel missionary Ramseyer recounted the frequent consumption of maize while being transported by the Asante military from the (present-day) Volta Region to Kumasi in 1869: "The next morning was brilliant, but our early walk through dripping foliage

[20] Huydecoper, *Diary*, 74.

[21] Perregaux, 'Chez les Achanti', 19.

[22] McCann, *Maize and Grace*, 49.

drenched us completely. After partaking of a small maize loaf, we pursued our way through tangled woods, treading the marshy soil with grateful appreciation of the shade afforded by trees."[23] This account was testimony to the portability of maize, which enabled the army to move quickly from region to region to subdue rebellions or capture territory.

By the nineteenth century, the ubiquitous maize-based coastal food *kenkey* had made its way into the forested interior. This bread was made of ground, soured maize, steamed in plantain leaves and served with soup, and was universally popular. Emissaries in Joseph Dupuis' party were presented with humble "cankay" bread by the Asantehene in an elaborate ceremony.[24] William Hutton recalled the consumption of kenkey in the forest settlements in Asante: "The natives of Paintrey live chiefly upon plantains, yams, cankey, and dried sea-fish, which they get from the coast…soup is also made from palm-nut, and is a common dish."[25] Kenkey's popularity in the forested interior was a testament to the extent and importance of intra-regional food trade and exchange within Ghana.

In Bowdich's detailed account of his journey from Cape Coast, he observed food production for the capital at approximately 60 miles to the southwest of Kumasi. In rings of small villages or *crooms*, he remarked that the rich black loam of the soil-sustained plantations of maize, yams, ground nuts, terraboys (runner beans), and *encruma* (okra). While in this area, he observed a fishing weir constructed out of wicker work and tree trunks, saw an excellent pig nursery, and commented on blacksmith and pottery production. He stated that yams and groundnuts were planted in raised triangular beds (to preserve moisture), with small drains around each, and were carefully cleared of weeds.[26] Many of the new starchy vegetables like maize were intercropped with yams and could be prepared quickly by simply boiling and serving them with soups or stews made from okra, groundnuts or peanuts.

---

[23] F. A. Ramseyer and Johannes Kühne, *Four Years in Ashantee* (London: James Nisbet & Co., 1875), 57.

[24] Joseph Dupuis, *Journal of a Residence in Ashantee* (London: H. Colburn, 1824), 103–04.

[25] Hutton, *Voyage to Africa*, 155.

[26] Bowdich, *Mission*, 31.

Kwabre (38 km northeast of Kumasi) and Atwoma (just south of Kumasi), were the principal settlement areas for slaves captured in battle since the early eighteenth century. These locations were used to grow food for the capital, including its communities of foreigners.[27] Many paths branching off from the main road were well travelled with strong evidence of the exchange of produce and manufacture.

> The extent and order of the Ashantee plantations surprised us…Their neatness and method have been already noticed in our route up. They use no implement but the hoe. They have two crops of corn a year, plant their yams at Christmas, and dig them early in September. The latter plantations had much the appearance of a hop-garden well fenced in, and regularly planted in lines, with a broad walk around, and a hut at each wicker-gate, where a slave and his family resided to protect the plantation.[28]

The farms that ringed the capital existed solely to provision Kumasi, for the benefit of the state apparatus. Part of the surplus was used in diplomatic efforts and lavished extensively on visitors.[29]

As discussed in Chap. 1, in the mid-nineteenth century, farmers working the land outside of the capital of Kumasi began to add considerable amounts of cassava to their crop rotations due to declining soil fertility, wrought by at least two centuries of intensive cultivation. Confronted with increasingly abbreviated fallow periods, farmers north of Kumasi (Kwabre) and to the southwest (Atwoma) found that cassava flourished even on degraded soils.[30] One of the most popular ways to prepare cassava was as fufu or as a different sort of dumpling called *kokonte*. The cassava was washed, peeled, and cubed. After being spread out to cure in the sun for a week, it was pounded in a mortar and sifted. The fine powder from the sifting was put into salted boiling water and stirred with a stick until it had the consistency of dough. Then it was rolled into balls and served with soup.[31]

---

[27] La Fleur, *Fusion Foodways*, 170.

[28] Bowdich, *Mission*, 325.

[29] McCaskie, *State and Society*, 26–37.

[30] Boaten, 'Asante', 95; Wilks, *Forests of Gold*, 52–53.

[31] Field, 'Gold Coast Food', 7–19.

While the local production of yam, maize, plantain, cassava and cocoyam were critical to Asante growth and expansion in the first half of the eighteenth century, consumption of imported foods, such as rice from Axim on western Gold Coast, rose in importance as a symbol of status by the nineteenth century. Wesleyan missionary Thomas Birch Freeman on his 1840 journey to Kumasi was gifted rice among other items by the Asantehene's aunt, Seiwa the Queen of Dwaben, 28 km east of Kumasi.[32] Indeed, rice was taken by the Asantehene and his officials at their main meal of the day at 2 pm, along with European biscuits, tea, and wine.[33] African rice had to be transported from the coast and took a great deal of labour to process. Henry Meredith, governor of Winneba fort in 1812, spoke highly of the rice produced in Apollonia (now part of the Western Region of present-day Ghana), and how the many rivers made the area suitable for rice and sugar cane cultivation.[34] Ghana's rice growing strategies were extremely localised. The Asante could obtain rice from the southwest rainforests near the coast, or from marshy areas near rivers like those in the Volta region. Farmers in the uplifted southern edge of the Volta River Basin on the Kwahu Plateau used highland varieties and practiced flood rice-field techniques. Lowland varieties were planted in evergreen rainforests of the southwest in rice gardens, where poorly drained soils would not support any other crop.[35]

General low meat consumption for people outside the court was documented by William Hutton, who accompanied Joseph Dupuis on the diplomatic mission to Asante in 1820. Ever attentive to the material culture of the time, he observed: "They have besides plenty of poultry, sheep, and hogs, but do not commonly indulge in this sort of food, as it is too expensive, and they prefer reserving these articles to sell to any principal men, either in the town, or to travellers."[36] A forest product obtained through gathering, snails as a source of protein, especially for poorer people, featured prominently in the cuisine, and continues to be part of the

[32] Thomas Birch Freeman, *Journal of Various Visits to the Kingdoms of Ashanti, Aku, and Dahomi, in Western Africa* (London: John Mason, 1844), 161.

[33] McCaskie, *State and Society*, 34.

[34] Meredith, *Account*, 45.

[35] La Fleur, *Fusion Foodways*, 98.

[36] Hutton, *Voyage to Africa*, 156.

fare today. An important source of income when sold at market, they were gathered at the beginning of the rains and sold smoked on a skewer.[37] The proprietors of the land required a substantial share of the produce of the land, which included snails (along with fish and game) be surrendered as payment from those permitted to settle the land.[38] Snails were considered such an important and marketable protein in the early twentieth century that colonial boundaries were calculated with reference to them. If a chief could prove that his men were in the habit of gathering snails over a certain area, this was accepted as proof that he held jurisdiction over that land. They were a frequent and important enough of a source of protein that they were included in avoidances and taboos on one's *ntoro* or patrilineage.[39]

Palm oil and its complementary relationship with dried fish from the coast, popularly known as "stink fish", was foundational to the taste and consistency of any dish. Palm oil was *the* vegetable oil, foundational to cuisines in both the forested centre and on the coast, where it was widely exchanged for salt. Highly prized stink fish, much like garum—a fermented fish sauce used as a condiment in the cuisines of the ancient Mediterranean World—provided a rich umami flavour to dishes.[40] Processed on the coast, stink fish was a convenience food of sorts, a precursor to the use of seasoning cubes today. Much like olive oil was served as a complement to garum, palm oil was the perfect complement to stink fish, providing additional richness, colour, flavour, and texture.

Diplomatic Food Gifting and Exchange

In the early nineteenth century, European relations with Asante were dictated by the abolition of the slave trade, and by a sequence of Asante military invasions of the Central Gold Coast in 1806, 1811 and 1816. It was at this time that the British became fully aware of the political and

[37] Allan Wolsey Cardinall, *In Ashanti and Beyond* (London: Seeley, Service and Company Limited, 1927), 78–80; Bowdich, *Mission*, 324; Hutton, *Voyage*, 329.

[38] Wilks, *Forests of Gold*, 99.

[39] McCaskie, *State and Society*, 171.

[40] David Downie, 'A Roman Anchovy's Tale', *Gastronomica* 3, no. 2 (2003): 25–28.

military power of the Asante. The Asante's partial absorption of the northern Fante country in 1816 and their demands for the rents ("notes") on coastal trading forts and castles that were formerly paid to this population of Fante marked the beginning of a new era on the Gold Coast.[41] The situation was tense due to the continued expansion of Asante power into coastal hinterlands of the European factories. The Danish had joined with the British, while the Dutch remained neutral, thus affording them the bulk of the gold and ivory trade from the interior.

The system of great roads in greater Asante in the nineteenth century consisted of a ring road with Kumasi at the centre, and eight trunk roads that radiated out: four heading north and four heading south. Traffic on the roads was controlled with a combination of surveillance by wardens, or *nquam-saraf* and sanctioned good and bad travel days.[42] This allowed the Asante to limit the flow of weapons to the north—an important source of slave labour—and to control trade beyond the northern region of their empire. At the same time, the Asante could control the movement of trade and information to the coast.[43]

With the European and local inhabitants of the coast finding that the cessation of the Atlantic slave trade upended the local economy, relations with the towns surrounding the forts became volatile and it was increasingly difficult to maintain order. Robberies increased in frequency. Governor Meredith of Winneba was killed in 1812 over abolition-related discontent. In 1815, the Dutch Governor, Daendels, cooperated with the British in a three-pronged strategy of pacification that included education, Christianity, and espionage.[44] Daendels had large scale plans: to open up communication with the powerful empires of Borno and Congo, and with the cities of Timbuktu and the Hausa region; to reveal the secrets of the source and the mouth of the Niger; and to employ the resulting knowledge in bringing about the increase of commerce, taking it away from the caravans of Tripoli. His overreaching goal was to open

[41] Edmund Collins, 'The Panic Element in Nineteenth-Century British Relations with Ashanti', *Transactions of the Historical Society of Ghana* 5, no. 2 (1962): 79.

[42] R. S. Rattray, *Ashanti Law and Constitution* (Oxford: Clarendon Press, 1929), 111.

[43] Wilks, *Asante in the Nineteenth Century*, 20, 37, 270.

[44] Collins, 'Panic Element', 99.

these markets to gain an estimated 100 million new customers for his trade goods.[45]

With this stated objective, Deandels surprised the British by deploying Huydecoper to Kumasi to begin negotiations for a treaty with the Asante. What follows is an examination of the experiences of Huydecoper in 1816, British envoy Thomas Edward Bowdich in 1817, Joseph Dupuis in 1820, Methodist missionary Thomas Birch Freeman in 1839, and Governor Winniett in 1848. Their encounters with Asante diplomacy, especially regarding food and its role in the expression of Asante power and style, are examined in the following case studies.

While the Asante employed military force in the regulation of trade routes to establish and maintain regional hegemony, foodways played an overlooked role in Asante's political dominance. Asante statecraft and its symbolic use of food was apparent in their diplomatic approach. Food was a vital diplomatic currency used by the Asantehene to convey favour, to express displeasure, to manipulate, to maintain hegemony, and to demonstrate power and competency. Food was used in a diplomatic context to produce and reproduce power relations in the nineteenth century.

Food gift-giving may seem a voluntary, innocent, and non-partisan act when it, in fact, it is laden with compulsion based upon economic self-interest. Marcel Mauss observed that hierarchy is established by means of gifts; to give something is to show one's superiority. Conversely, to accept gifts without return or repayment is tantamount to subordination in the relationship.[46] Furthermore, he observed that the economic practices of societies centred on the obligations to give, to receive, and most importantly, to reciprocate. This reciprocation circulated wealth and goods as well as built solidarity among those involved. The Asante were using this economic principle to manipulate their relationships with their neighbours, defining their connection with food gifts that Europeans would find difficult to repay. This in turn created an affiliation whereby the Europeans were considered clients and the Asante conceptualised themselves as magister. The following case studies allow an historical insight

[45] H.W. Daendels, *Journal and Correspondence of General H. W. Daendels* (Legon: Institute of African Studies, 1964).

[46] Mauss, *The Gift*, 1–2.

into the workings of the Asante state through the lens of food, feasting, and gifting.

Huydecoper

Willem Huydecoper made the journey to Kumasi as envoy at the age of 28. He diligently listed the "presents" given to him by the Asantehene and his generals, *caboceers* (captains), and their wives, and where they were received. The mostly food-related gifts given to him by the Asantehene and his generals were revealing. The most common items were bunches of plantains (671), yams (331), and sheep (15).[47] Much of the gold received was given to him for the express purpose of his entourage's upkeep. En route to Kumasi, he was a guest in General Appiah's camp at Abra and on 1 May he wrote: "No news today. I have enough provisions to last my men 4 days. About 12 o'clock I received two trays of food from General Appiah, and from the 'great officer' Cudjo Apannij 8 small kantjets [folded packets of gold dust] and one [illegible] of gold to buy meat. I was in no position to refuse these gifts and passed them on to my men."[48] The next day Huydecoper reciprocated with "two fine fowls baked in the pan", to which General Appiah sent palm wine and food, much to Huydecoper's delight. This pattern of the exchange of food gifts continued for his entire residency.

In Kumasi, food was used to bribe and/or induce obligation. Huydecoper had an earnest conversation with General Yaw Kokroko after he came to pay his compliments. Kokroko did not receive a flag from the Dutch governor as the other generals had. He had neglected to send his envoys with sword or staff to Elmina. This meant that General Daendels had not known whether his station required a gift. Kokroko brought Huydecoper a bunch of plantains and two chickens as encouragement. Huydecoper prevailed upon him to be patient as he was awaiting envoys who had been sent to Elmina to bring more gifts for the court.

[47] Huydecoper, *Diary*, 88–90.

[48] Ibid., 4.

On the *Kwasi-adae* ceremony of the Akan calendar, otherwise known as good Sunday, Huydecoper was given three British gold pieces by General Appiah with which he purchased a sheep for himself and his men. After the assembly, Appiah urged him to visit the other generals, Barriekie, Amanqua, and Yaw Kokroko to bid them good morning. He received various edibles such as sheep and bananas (plantains). He remarked, "All this visiting reminded me of New Year's Day at Elmina, when children run from house to house collecting their presents. To tell the truth, I felt more than a little ashamed."[49] The feeling of shame induced by what was essentially supplicatory collection of food from high-ranking cabinet members, was intentional and part of the diplomatic apparatus for any envoy that came to Kumasi.

The Asantehene withheld food to express displeasure directed at Fante envoys from Elmina who had defied him by, among other things, not turning over the rents they charged Europeans to occupy their coastal forts. The Elminans were there to drive a rift between the Dutch Governor and the Asantehene as these envoys stated that general Daendels was conspiring with the Wassaws and with the British at Cape Coast Castle to build a powerful fortress at Komenda and that Huydecoper's true purpose was to spy on the Asantehene to assess his strength. Huydecoper successfully convinced the Asantehene that the Elminans were lying by declaring his intention to drink a purification oath to prove his innocence and that of his master. This seemed effective as Huydcoper later wrote: "The Elminas are being abused on all sides because of their attempt to bring me into disfavour with the King and embroil him with the General…Today the Elminas received their gifts from the King. These were a cow, 21lbs 10 [indecipherable], 40 bunches of bananas (plantains) and 30 yams. There are 20 of these Elminas in all."[50] His tone indicated that this was not nearly enough to sustain this large a group of people. The next day, the Elminas killed their cow and sent a piece to Huydecoper, which he gave to his men. Seemingly, the Elminas were interested in currying favour with the Dutch envoy after their defamation attempt. The Elminas continued to suffer from hunger for the duration of their stay,

[49] Ibid., 8.
[50] Ibid., 39.

suffering displeasure from the King and comments from him that he "recognises no ruler in Elmina but the one who supplies him with his powder, guns, drink, etc."[51]

During his long residency in the court of Osei Tutu Kwame (May 1816 to April 1817), Huydecoper often encountered difficulty feeding himself and his men. On 29 August he wrote:

> At midday General Youw Kokroko visited me and saw me sitting miserably at table, dining most humbly off a plate of choking malaget pate. He was very sad to see me thus, and asked if I sat like this every day. Hearing that I did, he went home and sent me a sheep and 12 yams, so that I could have a better dinner tomorrow. I was much touched by his friendship, which far surpasses that of other persons here.[52]

Here Huydecoper was referring to fried plantains seasoned with malagueta, also known as "Grains of Paradise" or *Aframomum melegueta*, the seeds of a spice related to the cardamom family. Although a satisfying dish when accompanied by roasted groundnuts or peanuts, many would consider his almost exclusive consumption of plantain as a sign of poverty or of lowered status. Indeed, Huydecoper recorded difficulty in feeding himself again on 28 February when he was "reduced to eating fried plantains." He lamented that he could not even borrow money to eat, but frequently witnessed envoys laden with presents for the Asantehene from Governor Daendels coming into Kumasi. By March, the situation had worsened even more as plantains "cost three tackoes a bunch," takoes being a denomination of gold. Eventually, the King did provide the Dutch envoy with gold for his upkeep. On 1 May, Huydecoper received eight small *kantjets* with which to buy meat. On 5 May, he received three British gold pieces to buy a sheep. On 5 January, he was gifted with nine British gold pieces for his subsistence, and again the same amount thirty days later. Nevertheless, he still experienced issues with feeding himself and his men. In one example, on 11 May, he complained of being forced

[51] Ibid., 40.
[52] Ibid., 50.

to give his men gold with which to purchase food, due to its increasing cost.

Having control of food circulation in the capital and an awareness of the precarious food security situation beyond, owing to the ongoing conflict with the rebellious Fantes, the Asantehene was quick to reward the envoys with food when he perceived he was getting a desired result in negotiations. In November, six months after the start of the residency, the King made it clear to Huydecoper that he could not leave until all outstanding palavers (i.e. negotiations) had been settled. He pressured Huydecoper to deceive Governor Daendels about the Elmina's attempts to defame the Governor, fearing that he would retaliate by firing the town, depriving the Asantehene of income, and shaming him for being unable to control his vassals. After days of haranguing, Huydecoper finally relented, reasoning that the chief object of his mission would be compromised if he did not agree to this deception. The King rewarded him with food: "Everything was now done, and the King sent for a large bottle of rum and a fat sheep for our sustenance, saying that these were to show that we were friends and would remain so."[53] This reward of food came after months of food insecurity for the Dutch envoy in Kumasi. The control of the food in and around the capital was one of several ways that the Asantehene displayed is status and maintained order.

Another way the Asantehene displayed his wealth and power was the adoption of an opulent European-style table service that required maintenance. Near the end of Huydecoper's visit, the King requested that he take some young boys with him to learn the elements of European dining etiquette and style. "Today the king placed three boys in my charge. Two are destined to be carpenters and one a smith. His Majesty said he would later send more boys who would serve the General at table, and learn how to clean silver, glassware, plates, etc." The Asantehene's preoccupation with dining style was evident in his concern over maintaining and increasing the number and variety of implements of his European table service. On no less than three occasions, enquiries were made directly from the Asantehene to the Dutch envoy about the care of, and his desire for, the acquisition of more tableware. On 24 May, the King showed Huydecoper

[53] Ibid., 60.

his silverware collection and asked, "why such beautiful things were not sent to him anymore." Huydecoper told him of the war in Europe which was preventing the trade for such items and assured him that a silver cup was being sent for, at the request of his Asante envoys in Elmina. On his first and unexpected official visit from the Asantehene on 2 June, the King asked him to accompany him, whereupon reaching the palace, "he gave me all his silver ware and asked me to have it cleaned. This I agreed to arrange."[54] On 18 June, the King sent for four forks and a large salver to be cleaned while he was visiting the envoy. On 25 March, Huydecoper sent a letter to General Daendels in which the King requested silver utensils such as soup bowls and large basins in exchange for the cotton that was en route to Elmina.

A good deal of precolonial consumption in West Africa consisted of tributes or gifts. This was especially true of the Asante empire in the nineteenth century, which was in the process of becoming increasingly centralised.[55] The ritual and political centre of an empire encompassing many peoples and states, the Asantehene was conscious of the proper projection of his person. He perceived silver tableware to be an external symbol of power and used it unremittingly in order to impress his magnificence upon foreign representatives, as well as upon his own people.[56] He correctly understood that for the British, status came to reside in the ownership of objects, not people. By displaying opulent tableware of European manufacture, he was relating to European ideals of refinement and virtue. At the same time, he was tapping into the ritual of sociability that tableware came to represent for Europeans. Maintaining existing (and requesting additional) gifts of tableware enabled the Asantehene to display his character as one of respectability, conviviality, politeness, and power.

---

[54] Ibid., 61.

[55] William Tordoff, 'The Ashanti Confederacy', *Journal of African History* 3, no. 3 (1962): 399–417.

[56] T. C. McCaskie, 'Innovational Eclecticism: The Asante Empire and Europe in the Nineteenth Century', *Comparative Studies in Society and History* 14, no. 1 (1972): 30–45.

Bowdich

Bowdich's account of his party's official reception in Kumasi in 1817 has been widely used by scholars to make observations about Asante culture and customs. Fiona Sheales' ethnographical approach used material history to examine displays of diplomacy in the Anglo-Asante relationship. She noted that during official state receptions, envoys were subjected to sensory manipulations that incapacitated their ability to act. Asante hosts deliberately withheld food and fluids from envoys in temperatures that exceeded 30°C for periods in excess of five hours.[57] The absence of offers of food and water at official receptions suggested a conscious effort to manipulate envoys and reinforced their powerlessness in Kumasi.

Like Huydecoper before him, the British envoy's goal was to create a treaty and to establish a road to the coast that would allow the Asante to bypass the Fante and trade directly with the British.[58] Also like Huydecoper, British envoys received numerous gifts of food to create feelings of good-will and to obligate the envoys to the Asantehene and his military staff. On 24 May 1817, Bowdich wrote of an audience whereby Messrs James (the then commander of the mission), Tedlie, Hale, and himself were summoned to the palace of the king. The issue at hand was the handover of the "notes" or rents the British paid the Fante, who had recently been defeated in battle by the Asante. These notes passed to the Asantehene in 1816 when, after the revolts having been subdued, the Fante sued for peace, granting the Asantehene the notes for these castles. The day before this audience, the king presented the group with generous offerings of food, including a bullock and two sheep, in addition to billets of wood, yams, plantains sugar cane and other goods. Presumably, the Asantehene was preparing the ground by presenting the party with gifts to obtain a favourable result in negotiations for the rents he was now entitled to.

Initially, Mr James could not promise the Asantehene the missing rents, citing that he did not possess the requisite authority to ensure this outcome. This enraged the Asantehene, whose retinue drew swords and

---

[57] Sheales, 'Sights/Sites of Spectacle', 93.

[58] Collins, 'Panic Element', 99.

swore to take the fight to the Fantes. The party was sent away and summoned again two hours later, where the King explained how the pittance he was receiving from the Fante for rents was shaming him. Claiming he saw the dire situation they were in, Bowdich took matters into his own hands, stating that "We were sent to make the British and Ashantee as one", whereupon he swore upon his sword, as did Tedlie and Hutchinson. This action seemed to bring the King some relief. His linguist made the following proclamation to Bowdich:

> When you three white men go back to Cape Coast, and the Governor has bad put into his head, and thinks you did wrong, then if you want anything to eat, send a messenger to him and he will send you plenty, for the King thinks you do right to God and him, and to your King and to the Governor, and that you will get much honour when you go back so the King thanks you, and says you speak well.[59]

The Asantehene's linguist went on to say that if the British would place their trust in him, he would send them all his trade, and good gold (not bad gold he knew the Fante sent), and that "If at any time the British in the forts are in want of anything to eat, and send to him, he will send them everything."[60] The King offered a measure of food security as a reward for a favourable diplomatic solution to a contentious problem.

Bowdich's account of the way his party sustained themselves differs from Huydecoper's in that Huydecoper, albeit with some reservations, relied upon his hosts for sustenance and actively requested food or gold in order to purchase food from the King. Bowdich's party all along had been receiving gifts of sheep and fruit every few days. He remarked that the trail to Kumasi had been cleared for his party by order of the king, and that his party periodically received gifts of prestige food and gold on the trail via messenger.

> We passed Datiasoo, where large quantities of pottery were manufactured, exclusively: it was not more than a mile distant from Dadawasee, where we found a messenger from the king, expressing his regret that we had come

[59] Bowdich, *Mission*, 63.
[60] Ibid., 64.

> up in the rainy season, as he had heard it was a very unhealthy one for white men, and appointing us to enter the capital on the Monday following; he sent us a present of a sheep, forty yams, and two ounces of gold for our table.[61]

The delivery of prestige foods like meat and yam on the voyage to Kumasi and the generous distribution of gold was meant to communicate Asante power and wealth to the British. When contrasted with the relatively modest food gifts Huydecoper received, the argument could be made that the British may have been perceived as more desirable trading partners than the Dutch. It is possible that this difference in treatment was reflective of the growing power of the British relative to the Dutch on the coast during this period.

The gifts bestowed upon Bowdich's party after they were installed in Kumasi on 17 June, seemed to reflect the standing of the British in the Asantehene's valuation:

> 2 ounces of gold to the officers.
> 20 ackies to our people.
> 10 ackies to our linguists.
> 1 hog, 1 sheep, and a profusion of plantains and oranges.[62]

Bowdich wrote his interpretation of this gift as a "reproof of a disgraceful attempt to borrow money of him for our subsistence", stating that he considered this type of behaviour as beneath his dignity. This attitude on the part of Bowdich reflected the tensions between the British envoy's elevated perception of himself vis-à-vis the Asantehene's perception of him and the British as less powerful. By lavishing gold on the envoys for their sustenance in Kumasi, the Asantehene successfully demonstrated that he occupied a superior position, a perception that caused Bowdich unease.

The official state dinner on Monday 25 August was announced at 2 o'clock. Bowdich stated, "We were taught to prepare for a surprise, but it

[61] Ibid., 31.
[62] Ibid., 83.

was exceeded."[63] This was after the day started with a sumptuous breakfast in a purpose-built edifice. Expertly cooked soups, stews, plantains, and rice were served at breakfast, along with wine and fruit. Bowdich noted that the servants and messengers were "distinctly provided for". The dinner venue was in the king's garden, described as about the size of one of London's larger squares. Four large scarlet state umbrellas protected the king's elevated feast table, giving the occasion a feeling of solemnity and pomp. The Asantehene did not sit with his guests but a short distance away with his captains. He visited his guests frequently, conversing and enjoying their ribald toasts. The Asantehene's decision to not sit with his guests projected distance, separation, and formality. Indeed, the Asantehene customarily took his meals alone—a tradition he was not willing to break with for these guests. Tables were elevated, and laden with heavy plates and with both European and local food. Native soups, fruits, dessert, and spirits were served on either side of the elevated centre table on the ground or on low tables. The efforts of the Asantehene to project a sense of majesty with the picture of overflowing abundance, the adoption of European table service, and European social dining etiquette proved successful. His guests were suitably impressed. Bowdich related "We never saw a dinner more handsomely served, and never ate a better. On expressing our relish, the King sent for his cooks, and gave them 10 ackies (gold)."[64] He retired after the feast to allow his servants to clear the table. The entire dinner lasted approximately three hours, and at the end of the dinner the King gifted the British servants the remainder of the wine and cordials, the tablecloths and napkins, and sent them all home with cold pig and fowls for their supper. Although he did generously provide for the requirements of his guests and their servants, the King did not spend much time getting to know this party on a personal level. Ever the gracious host, he presided briefly and then ceremoniously withdrew.

[63] Ibid., 114.

[64] Ibid., 130.

Dupuis

The treaty Bowdich negotiated with the Asantehene was not ratified due to complications over the notes for Cape Coast Castle and Anomabo Fort. Consequently, Joseph Dupuis was sent in 1820 to renegotiate. In addition to negotiations over the fort notes, Dupuis' mission was to collect information on Asante history, geography, and of neighbouring countries in the interior. Dupuis was also to make inroads toward securing better terms of trade with the king from the regions in the interior. He left Cape Coast on 9 February 1820 and did not return until 5 April. The treaty itself was signed on 23 March 1820.

On the journey to Kumasi, the King sent two messengers while Dupuis' contingent was en route. The first visit from a royal messenger on 23 February saw Dupuis witnessing two bearers carrying "a fat sow" all the way from Kumasi, along with a sheep and gold dust encased in silk given for their sustenance. That same day, a second royal messenger brought the king's own six-year-old son, along with his young attendants, to meet Dupuis' party near the village of Sanquanta. This party brought presents of fowls, vegetables, and wine to complement the earlier gift.[65] This exchange was a display of Osei Tutu Kwame's dominance, his largesse, and his complete confidence in the lack of any threat from the British. The extravagance of having bearers carry a fat sow all the way from Kumasi instead of herding the animal, was an expression of wealth and power in the form of food.

During Dupuis' residency, an interesting pattern emerged whereby the Asantehene would send a small gift of food to invite Dupuis to attend him in court in advance of negotiations, or of other gift giving occasions. On 4 March, shortly after his arrival in Kumasi on 28 February, the King sent Dupuis a gift of palm wine and a message that he should attend him in the "great area…where he designed to distribute some bulky presents before the people".[66] Loath to attend such a public spectacle, but encouraged by the Asantehene's Muslim courtiers, Dupuis was escorted to the appointed place by a guard of honour. Dupuis was seated with the other

[65] Dupuis, *Journal*, 58–61.
[66] Ibid., 101–02.

officers of the court, along with a retinue of Fantes, in front of a full assemblage of the people. The preliminary gift of gold dust was distributed and after a pause, with the joyous gaze of the Asantehene trained upon him, Dupuis and the Fantes with whom he shared quarters, were treated to the most extravagant food gift recorded by a European envoy to date.

> Soon afterwards a file of about two hundred men and boys, laden with provisions, passed in the order of review, and were then conducted to my quarters, where they deposited their burthens. The supply consisted of two cows, several sheep and goats, poultry, eggs, yams, plantains, bananas, honey, oil, palm, nuts, and wine, cankay bread, and a large pig…Upon approaching the avenue leading to my apartments, the scattered heaps of provisions which had been cast promiscuously on the ground, nearly choked up the way.[67]

Dupuis reported that this "princely donation" did not last as expected. Along with their own daily consumption, the townspeople pilfered the bread, wine and nuts. However, the shortage of livestock in the tsetse zone, where sleeping sickness took a heavy toll on domestic livestock, meant that gifts of livestock were largely supplied from the northern savanna region. Therefore, each gift of livestock spoke of Asante domination of trade routes to these regions, and a diet rich with domesticated meat, whereas the food of the "poorer classes" was composed of bush meat such as deer and monkey.[68]

Dupuis' subsequent remark about the Fante revealed the Asantehene used food deprivation as a diplomatic tool. Upon the last day of "Little Adai" on 15 March, seated with the king's captains under a canopy, Dupuis received his portion of the gifts being distributed on that day: "and my share of the bounty was a fat sheep, and a bottle of rum. Even the Fantees received for their portion four sheep and two kegs of rum".[69] The Fantees were evidently maintaining their residency in Kumasi for several years in the hope of repairing the relationship between themselves

[67] Ibid., 103–04.

[68] Bowdich, *Mission*, 319.

[69] Ibid., 142–43.

and the Asante. Judging by the quantity of food gifts the Fantee received, one might infer that the esteem with which the Asantehene held them in had improved since Bowdich's visit three years before.

Freeman

By the time British Wesleyan Methodist missionary Thomas Birch Freeman undertook his first mission to Kumasi in 1839, the local political climate had changed. Freeman was born in Hampshire in 1809 to a British mother and a freed African slave. A dedicated man of faith, he lost his post as head gardener on a Suffolk estate due to his Methodist activism.[70] Freeman served as escort for two of the Asantehene's nephews who were being educated in England. This case study shows the contrast between the approach of Osei Tutu Kwame in years past, and Asantehene Kwaku Dua I (1834–1867). The changed political reality Freeman encountered was due in part to the first Asante military defeat, which took place at Katamanso on the Accra plains on 7 August 1826.

In standard diplomatic procedure, while journeying from the Cape Coast Castle to Kumasi and stopping at Korinchi, Freeman received a gift of gold from the King via messenger for the maintenance of his retinue on the journey.[71] According to Freeman's account of his first visit, the official reception only lasted one and a half hours, and he was refreshed: "While I was sitting to receive the compliments of some of the first chiefs who passed, His Majesty made me a present of some palm wine."[72] Similar to the experience of Huydecoper, Freeman's encounter included the receipt of gifts on days that were of particular significance to him, as a person acculturated to European holidays. "At two, the King sent us a present, consisting of two fat sheep; one for the Princes, and the other for Mr. Brooking and myself. Apoko stated, that His Majesty had heard that this was Christmas-day, and that he sent the sheep by way of

[70] 'Freeman, Thomas Birch (1809–1890)', Boston University School of Theology History of Missiology, accessed 3 March 2016, http://www.bu.edu/missiology/missionary-biography/e-f/freeman-thomas-birch-1809-1890/.

[71] Freeman, *Journal*, 22.

[72] Ibid., 48.

congratulations."[73] He followed up this gift an hour and a half later with some palm wine. Ensuring he was kept abreast of British custom, the Asantehene demonstrated his cosmopolitanism by sending a gift of food for Freeman's party to use in their holiday celebrations.

At seven in the morning on 28 December, Freeman and his party received an invitation to a state dinner. The King made an unusual request: that Freeman loan the King his cook to assist with the dinner's preparation. The reason for this request was to become clear later in the day. Freeman's dinner, which commenced at 2:45 pm, included kidney beans "well served up in the European style", roast fowl, roast mutton, and fish. In contrary to the dictates of ritual that demanded the Asantehene take dinner alone, the king ate with his guests. It became apparent then that the Asantehene requested the loan of Freeman's cook so that his own cook could learn how to prepare dishes in the English style. With his own cook preparing the food, the Asantehene's ritual purity would not be compromised.

The venue was in a large 80 by 45-foot yard, shaded by several state umbrellas. This may have been the very same space in which Bowdich was entertained. The Asantehene's table was lower to the ground and set with Portuguese silver plate. Freeman was placed at the head of the high table, flanked by the repatriated nephews of the Asantehene, William Quantamissah and John Ansah. The King sat at the head of the lower tables, facing Freeman. Instead of being presented with gifts as Bowdich's party was, Freeman requested they give thanks to God before they sat down to dinner, a request to which the Asantehene happily acquiesced. The Asantehene was attended by linguists, the heir apparent, and another prince, all of whom sat near the table. The captains from the King's own house sat behind the princes. Most remarkably, using European instruments, the band played music familiar to Freeman. Asking a question that he well knew the answer to, the Asantehene inquired whether musical accompaniment was a feature of European dining. The king was highly conscious of the worldly impression he was making with this orchestrated display of European music and tableware.

[73] Ibid., 134.

The meal was European in style, yet it had interesting cross-culture eccentricities. Roasted sheep was served during the main course of the feast, and Freeman confirmed for the Asantehene that it was prepared in a similar way in Britain. The Asantehene cut the sheep with the assistance of his servants and sent some around to his guests. Freeman had some soup and a plum pudding prepared from his own stores. The reproduction of traditional British Christmas foods conveyed his knowledge of how high holy days were celebrated in England. The Asantehene gladly tasted these foreign foods as they were prepared by his own staff and sent them around to his entourage.[74] At the conclusion of dinner, Freeman drank to the Asantehene's health, the auspicious return of the princes, and to the hope that the Asante and the British would become one in spirit. After a fruit course in the courtyard, the guests were escorted home by the band, which played European tunes for them all the way back to their lodgings.

Freeman later reflected on the extraordinary events of the day, marvelling that the King of Asante, who holds the power of life and death in his smile or frown, sat at the table with Christian British men. These circumstances were extraordinary, especially when a comparison of commensality is made between the official dinner in Bowdich's account and that of the Freeman party. Commensality and its transmission of symbolic meaning defines relationships, status, and kinship. The choice to host the meal in the King's private courtyard in Kumasi—the inner sanctum of Asante power—was emblematic of the status of both parties in relation to each other. The official dinner in honour of Freeman, although seemingly in the same venue, seemed more intimate. Seating indicates the level of intimacy desired—socially intimate or distanced to either facilitate conversation or to keep the protocol by rank.[75] The layout of the table with the Asantehene eating with his guests and in the presence of his family members indicated the conference of equality in group status. By contrast, in the case of Bowdich's account, the two parties sat apart: a deliberate power play and an obstruction to conversation. At dinner with Freeman, the Asantehene cut meat for the main course with his own

[74] Ibid., 140.

[75] Morgan, 'Diplomatic Gastronomy', 146.

hand, signifying that he desired a closer connection with his guest while emphasising his status as host. Not to be overlooked are the two instances whereby the Asantehene tasted the British food he was offered. To eat food produced by another person is to experience that person both physiologically and emotionally. Food contains the self and the feelings of its producer, and the sharing of his guest's food and his guest's company at the table indicates the deliberate construction of social and political solidarity. [76]

Winniett

William Winniett was a Canadian Royal Navy officer who in 1845 became lieutenant governor of the Gold Coast. In his role as governor, his priorities were the promotion of the interests of British traders on the coast, the cessation of the practice of human sacrifice, and to impede the ongoing illegal slave trade. The first two segments of his mission led him to visit the Asantehene in September 1848. Accompanied by Rev. T. B. Freeman, Captain Powell of the First West India Regiment, and 150 hammock men and luggage carriers, on 4 October he and his party crossed the Prah River. Five days later they entered Kumasi, and they remained there until 26 October.

Winniett retired to the home of the resident Methodist missionary, Mr Hilliard, after his reception. Soon after he arrived, he received a gift from Asantehene Kwaku Dua that dwarfed even the gift that Dupuis received in 1820.

> 2 bullocks,
> 4 sheep,
> 4 turkeys,
> 6 ducks,
> 20 guineafowls,
> 6 pigs,
> 20 fowls,

[76] Anna Miegs, 'Food as a Cultural Construction', in *Food and Culture: A Reader*, ed. Carole Counihan and Penny van Esterik (New York: Routledge, 1997), 103.

20 pigeons,
400 yams,
303 bunches of plantain,
4 dishes native rice,
5 dishes ground nuts,
6 calabashes of honey,
oranges,
eggs,
palm-nuts,
sundry vegetables,
40 logs of wood,
40 baskets of corn.

This magnificent present was brought before me by 550 men, every one of whom had some share in the work of conveyance; these were accompanied by several officers of the King's household and their retinue, amounting to not less than 300 men, thus, about 850 men were employed in presenting to me this token of the King's good-will.[77]

This single gift almost equalled what Huydecoper received for his entire entourage over the year of his residency from Osei Bonsu.

At the banquet that followed, the tone was congenial. Pleading ignorance as to the use of cutlery, the Asantehene did not actually eat with Governor Winniett, but sat at the table across from him, took wine with him, and conversed with the Governor and the other gentlemen of his "suite". Dressed in an officer's uniform, the Asantehene presided over the serving of plum pudding and roasted sheep. The banquet ended with a visit to the apartments of the ladies of the court, an honour not afforded to even the Asantehene's closest chieftains. Rounding out the entire experience was a personal escort by the Asantehene, carried in his palanquin part of the way back to the mission house. Over the 18 days in residence, Winniett received gifts of sheep, fowl, eggs, yams, or plantain almost daily from the King. He interpreted these gifts as strong indicators of public feeling toward him as a representative of Queen Victoria.[78] It was

[77] PP 1849 XXXIV [C. 399] Missions to the King of Ashantee and Dohomey
[78] Ibid., 11.

more likely that these daily gifts of food and feasting were a projection of the Asantehene's own dominion and sense of grandeur.

Food and Power

Using the first-hand accounts from Huydecoper, Bowdich, Dupuis, Freeman, and Winniett, it is possible to construct a normative Asante court food culture. It is also possible to consider the change in style between Osei Tutu Kwame and Kwaku Dua Panin in their dealings with their European neighbours on the coast. Commensality made the Asante organisational structure obvious and allows for the construction of a normative palace food culture.

In each account, the Asantehene would leave after the banquet so that his servants could clear the tables. In the case of Freeman and Winniett, Kwaku Dua Panin would retire to receive his guests in another part of the palace complex in order to continue the evening's entertainment. Both Osei Tutu Kwame and Kwaku Dua took great care to ensure the servants of his guests were well fed, either at the time of the feast, or by sending food for the servants from the feast with his departing guests. At each banquet, the king's food was customarily served at a table low to the ground, and the display of elaborate silver plate was conspicuous.

Overall, these case studies make it clear that declining to eat with guests was part of the normative palace food culture. The fact that Kwaku Dua sat eating with Freeman and with his family members in attendance meant that this occasion was a singular honour. Observing Christian commensality customs, Kwaku Dua tasted the foreign foods as prepared by his staff:

> On sending the soup round, I asked, whether the King would take any, to which he answered, "Yes;" and when it was placed before him, he tasted it, and then, according to Asante custom, gave the remainder to some of his attendants who were near him. Osai Kujoh also tasted it. While we were

> taking our portion, a bountiful supply of native soup was placed before the King, which he sent round to his Captains and people.[79]

Kwaku Dua's approach to commensality was a marked contrast with Osei Bonsu's attempts to overwhelm his guests with conspicuous displays of giving before, during, and after banquets, and to every level of participant from the cooks to the carriers. Bonsu kept his distance during dinner and after the banquet as he did not escort his guests home. Osei Bonsu continued this more assertive approach in other contexts with Huydecoper, Bowdich, and Dupuis. Public food gift giving, lavish ceremonies, and leaving his guests Bowdich and Dupuis unrefreshed after lengthy observances all served to broadcast his power. This at times uncongenial approach reflected the political environment of Osei Bonsu's council, whereby he was under pressure to adopt more militaristic policies. His support of Dutch Governor Daendels was not popular with his council, who were persuaded that Daendels was an enemy of Asante. Osei Bonsu also had to contend with an abortive coup attempt by several royal wives and princes, whose conspirators he executed in 1819.[80] His distant manner could be indicative of ambivalence and anxiety over the internal power struggle and lack of support he was experiencing, manifesting itself through his diplomatic style.

Taking Freeman for a tour of the Stone House (built by Osei Tutu Kwame and housing a collection of the Asantehene's treasures) and Winniett for a tour of the ladies' apartments demonstrated a level of trust on the part of Kwaku Dua, and perhaps a desire to communicate his wish for an increased level of familiarity. The Asantehene physically brought himself close to his guests by sitting at the table, facing them when they dined, attentively catering to their personal food tastes, and making the choice to include his family. He shared his personal living space with his guests and ensured they had an escort after the banquet by providing the band in Freeman's case, or taking the time to escort his guests personally, as in Winniett's example. This new approach reflected the consolidation of the peace interests at this time due to the good relations maintained

[79] Freeman, *Journal*, 140.

[80] Wilks, *Asante in the Nineteenth Century*, 484–85.

with the British administration of Maclean. Ironically, Kwaku Dua ensured 34 years of peace by eliminating ringleaders of a plot to depose him led by the *Gyaasewahene* (Head of the Exchequer), Adu Dampte. During Freeman's 1839 visit, the *shaykh* of the Kumasi Muslims was detained for his role in the plot to depose Kwaku Dua.[81] Yet, his commitment to a peaceful administration is reflected in his approach to diplomacy.

The official dinners in all the accounts offer up important cultural signposts. Although the culture of cooking was a woman's domain, the Asantehemaa (queen mother), considered the most powerful woman in their matrilineal society, was responsible for the selection of the heir apparent. However, the Asantehemaa was not at the apex of the culture of cooking. That dual political and ritual role belonged to the Asantehene and his *soodoofoo* (male cooks) in their role as intercessors to their ancestors and gods. This was an important difference in social organisation that was observed in the official reception and during important events such as the Odwira.

The Asantehene had access to food sourced from different regions. For example, the bullocks he gave as gifts did not live in the forest zone, but further afield in the northern savanna. This reach was indicative of his access to and absolute control of trade with this region. The ability to gift and consume these animals with such regularity signified membership in the upper classes which dined regularly on meat, fowl, and fish from the coast. In Bowdich's account, most non-elites found it too expensive to slaughter animals regularly, relying instead upon a diet of starch and game. Eggs were another ritual food "forbidden by the fetish" to Asante, but freely gifted to European envoys.[82] The Asantehene signalled access to historical trade networks in the north with the gifting of animals. The argument can be made that the foods gifted to envoys reflected the hybrid cuisine of the elite in Kumasi, the centre of a powerful empire.

The Asantehene frequently surprised his guests with his access to foreign foods and style of presentation to include tablecloths, napkins, and silver plate. Access to foreign taste was power, and the elaborately

[81] Ibid., 488–89.

[82] Bowdich, *Mission*, 267.

presented foods and material culture from Europe marked Kumasi and the Asantehene as occupying the peak of wealth accumulation, political power and elegance. European preparation of foods such as roasted mutton, beans, and pease-pudding impressed those in whose honour the feast was being presented. Guests reacted with surprise at the presentation of sumptuous food and style in what was viewed as a remote forest kingdom, with items that not even they had regular access to. These dinners functioned as political theatre, conveying the domination of trade networks in the region.

Food was essential to the dominion held by the Asantehene, as explored by McCaskie in his analysis of hegemonic structures of power in the Asante state, and as performed during the Odwira. Theatrical display, gift exchange and control of food supply were vital components in the exertion of influence with subjects and outsiders alike. Under the framework of the power of prestige, manipulation of food supply, food gifting, and diplomatic ceremony served as a barometer for political relationships and as a mirror reflecting the power struggles both within Asante and outside of the state. Using "soft power" and cultural diplomacy to direct behaviour through perception, symbolism, and culture, the Asante mobilised food as a diplomatic tool. In so doing, Osei Bonsu achieved diplomatic goals, one of which was to win the rents or "notes" due to them from the Fante on the coast, and the stymying of direct trade of the Europeans with the northern savanna. Both Kwaku Dua and Osei Bonsu successfully executed diplomatic commensality, achieving their political objectives with the use of food.

Cooking changed from the fifteenth to the seventeenth century as individuals experienced greater convenience and speed of processing and preparation with the introduction, adoption, and incorporation of plantain, maize, and cassava. In addition to the benefits of faster processing and preparation, increased portability became a factor in the growth and expansion of the Asante empire, in that their armies were able to travel further and faster with the benefit of maize porridge and bread. In the nineteenth century, plantain and cocoyam were increasingly thought of as food for slaves, women, and children. Presumably, the ease of production in back-yard gardens and lack of social organisation needed to procure them, lowered their relative prestige to yams, which required intensive labour and represented the foundation of social relationships in Asante. Yams and rice remained prestige foods due to the time and expense of production and processing. By the end of the nineteenth and the beginning of the twentieth centuries, foods like plantain, cocoyam, and cassava became the food that immigrants, who came to the region to work primarily as miners, turned to.

The use of rice, plantain and cassava, however, did not transform cooking and recipes significantly. Rather, these foods were incorporated into existing dishes where similar seasonings were used and they could be processed to achieve desired textures. Plantain and cassava could be combined with yam to make fufu. In contrast, the introduction of maize did transform cooking and recipes, mainly in that fermentation was now a common practice in the cooks' repertoire. In fact, fermentation may have been the only significant change in food processing in the nineteenth century in Asante. Fermented maize and cassava aided in digestion and had increased nutritional value. Other significant changes would include the introduction of porridge as a staple dish in Asante foodways. Most dishes up to the introduction of maize were based upon a solid starchy mass served with pepper sauce, soup, or stew. Yam, for example, would be boiled and pounded into a glutinous thick dumpling to convert the starches, making them more easily digestible. With the increased use of both rice and maize, the addition of a thin maize porridge became another staple quick meal used as a breakfast or for travel.

Culinary changes were reflected in Asante diplomatic custom in those foods that required greater processing, such as yams and rice, were more highly esteemed. The treatment of Huydecoper provides a case study for demonstrating how both new and local food was used to communicate social standing in the Kumasi court. Most other envoys were gifted copious quantities of meat. Additionally, they were received with a state dinner and served with prestige foods like rice, yams, and imports. This use of prestige foods with British envoys may have reflected the perceived ascendancy of the British on the coast relative to the Dutch. Huydecoper, as a representative of the Dutch, was forced to eat an awful lot of plantain. The position of the Dutch at the time explains the disparate treatment of envoys in the Asante court. By the late 1780s, the financial position of the Dutch government in Elmina had become so precarious that the governor felt compelled to appeal to the local merchants to assist the government with huge loans. By 1790, the Dutch West India Company was bankrupt, and its possessions were subsequently taken over by the state, which continued the administration. The abolition of the slave trade by the Dutch in 1814, following the British abolition in 1807, forced Elmina Governor, Daendels, to experiment with new types of economic activities (such as plantations) to keep the enterprise viable.

One of Daendels' objectives was to start constructing a modern trade road to Kumasi, based on the Post Road he had built across Java several years before. In the field of diplomacy, a concerted effort was made to re-affirm the relationship with the Asante court by posting a resident representative (Huydecoper) in Kumasi. All efforts came to nothing, as Daendels died from yellow fever after only two years in Elmina, in May 1818. The Dutch government used the opportunity of Daendels' death to minimise their presence.[83] The Asante were assuredly aware of the precarious position of the Dutch on the Gold Coast and this perception was echoed in the grudging way Huydecoper was provided for by the royal court.

Rice, on the other hand, was a prestigious food consistently gifted from envoys to the court, and from the court to envoys from 1817 to the late 1840s. Governor John Hope Smith sent rice as a gift to the Asantehene along with Bowdich's party.[84] The Asantahene's aunt, and Queen of Dwaben, sent rice as well as yams, sheep, and nuts as a gift to Freeman.[85] The Asantehene ate yams and rice, while his family and friends tended to eat more plantain and maize. Retainers were sent to grow maize and plantain for the Asantahene's retinue and family on farms that ringed Kumasi. The elevated position of yams became codified in court ritual via the Odwira. Food was wielded as a powerful political tool, and that power was understood by Europeans coming to negotiate with the Asante in Kumasi.

---

[83] Ineke van Kessel, *Merchants, Missionaries & Migrants: 300 Years of Dutch-Ghanaian Relations* (Amsterdam: KIT Publishers, 2002), 26–27.

[84] Bowdich, *Mission*, 131.

[85] Freeman, *Journal*, 161.

Conclusion

The gendered characteristics of Asante foodways have been overlooked in the historical construction of their regional power. Food in Asante was highly gendered and subsequently weaponized against vassals and European trade envoys as an integral part of ritual and diplomatic spectacle throughout the nineteenth century. The control the Asantehene wielded over foodstuffs extended far and wide. In Kumasi the Asantehene controlled food production and distribution, reminding his subjects of his power at every meal. The gendered aspect of foodways in Asante was ritualized to command the fealty of vassals and to extract favourable trade terms with Europeans.

This finding builds upon the work of Ivor Wilks and James McCaskie on the introduction of plantain, maize, and cassava in the sixteenth century and their impact on the ability of the Asante to support a larger population, enabling them to increase their production of gold. A food studies approach to the study of the Asante provides a more nuanced view of their social history. It marks the change over time in commensality, gifting, and performance that reflected the decline of the relative power of the Asante in the region. Historical continuity is found in the desire for *fufu* which emanated from the ritual and symbolically gendered role of yam in the cosmology, which is extant to this day. The analysis suggests that the Asante enacted gender in ways that warrant closer inspection. This chapter puts forward new empirical findings about gender and food in West African history and employs the concept of "doing gender" to explore how masculinities and femininities were negotiated in Asante culture.

References

J.K. Adjaye, *Diplomacy and Diplomats in Nineteenth Century Asante* (University Press of America, Lanham, 1984)

A. Boaten, N. Akwasi, Asante: the perception and the utilization of the environment before the twentieth century. Inst. Afr. Stud. Res. Rev. **6**(2), 19–28 (1 January 1990)

Boston University School of Theology History of Missiology, Freeman, Thomas Birch (1809–1890), http://www.bu.edu/missiology/missionary-biography/e-f/freeman-thomas-birch-1809-1890/. Accessed 3 March 2016

T.E. Bowdich, *Mission from Cape Coast Castle to Ashantee* (J. Murray, London, 1819)

A.W. Cardinall, *In Ashanti and Beyond* (Seeley, Service and Company Limited, London, 1927)

E. Collins, The panic element in nineteenth-century British relations with Ashanti. Trans. Hist. Soc. Ghana **5**(2), 79–144 (1962)

D.G. Coursey, The origins and domestication of Yams in Africa, in *West African Culture Dynamics: Archaeological and Historical Perspectives*, ed. by B. K. Swartz, R. E. Dumett, (De Gruyter, Inc., Berlin, 1980)

H.W. Daendels, *Journal and Correspondence of General H. W. Daendels* (Institute of African Studies, Legon, 1964)

P. De Marees, *Description and Historical Account of the Gold Kingdom of Guinea* (British Academy, Oxford, 1987)

D. Downie, A Roman Anchovy's tale. Gastronomica **3**(2), 25–28 (2003). https://doi.org/10.1525/gfc.2003.3.2.25

J. Dupuis, *Journal of a Residence in Ashantee* (H. Colburn, London, 1824)

M.J. Field, Gold coast food. Petits Propos Culinaire (43), 7–19, (1994)

T.B. Freeman, Journal of Various Visits to the Kingdoms of Ashanti, Aku, and Dahomi, in Western Africa. J. Mason, 1844

W. Hutton, *A Voyage to Africa* (Longman, Hurst, Rees, Orme, and Brown, London, 1821)

W. Huydecoper, *Huydecoper's Diary: Journey from Elmina to Kumasi 28th April 1816–18th May 1817*. Translated by W. Graham (Legon: University of Ghana, 1962)

S. Kerner, C. Chou, Introduction, in *Commensality: From Everyday Food to Feast*, ed. by S. Kerner, C. Chou, M. Warmind, (Bloomsbury, London, 2015)

I. van Kessel, *Merchants, Missionaries & Migrants: 300 Years of Dutch-Ghanaian Relations* (KIT Publishers, Amsterdam, 2002)

J.D. La Fleur, *Fusion Foodways of Africa's Gold Coast in the Atlantic Era* (Brill, Leiden, 2012)

M. Mauss, *The Gift: Forms and Functions of Exchange in Archaic Societies* (Cohen & West, London, 1966)

J.C. McCann, *Maize and Grace: Africa's Encounter with a New World Crop, 1500–2000* (Harvard University Press, Cambridge, 2007)

T.C. McCaskie, Innovational eclecticism: the Asante empire and Europe in the nineteenth century. Comp. Stud. Soc. Hist. **14**(1), 30–45 (1972)

T. McCaskie, *State and Society in Pre-Colonial Asante* (Cambridge University Press, Cambridge, 1995)

H. Meredith, *An Account of the Gold Coast of Africa* (Longman, Hurst, Rees, Orme, and Brown, London, 1812)

A. Miegs, Food as a cultural construction, in *Food and Culture: A Reader*, ed. by C. Counihan, P. van Esterik, (Routledge, New York, 1997)

Missions to the King of Ashantee and Dohomey. House of commons papers. United Kingdom, Vol 34, (27 April 1849), https://parlipapers.proquest.com/parlipapers/docview/t70.d75.1849-026002/usgLogRstClick!!?accountid=16710

L. Morgan, Diplomatic gastronomy: style and power at the table. Food Foodways **20**(2), 146–166 (1 April 2012)

E. Perregaux, Chez Les Achanti. Société neuchâteloise de géographie **17**, 7–387 (1906)

F.A. Ramseyer, J. Kühne, *Four Years in Ashantee* (James Nisbet & Co., London, 1875)

R.S. Rattray, *Ashanti Proverbs: The Primitive Ethics of a Savage People* (Clarendon Press, Oxford, 1916)

R.S. Rattray, *Ashanti Law and Constitution* (Clarendon Press, Oxford, 1929)

C.C. Robertson, *Sharing the Same Bowl?: A Socioeconomic History of Women and Class in Accra, Ghana* (Indiana University Press, Bloomington, 1984)

F. Sheales, Sights/sites of spectacle: Anglo/Asante appropriations, diplomacy and displays of power 1816–1820. PhD diss., University of East Anglia, 2011

W. Tordoff, The Ashanti confederacy. J. Afr. Hist. **3**(3), 399–417 (1962)

I. Wilks, *Asante in the Nineteenth Century: The Structure and Evolution of a Political Order* (Cambridge University Press, London, 1975)

I. Wilks, *Forests of Gold: Essays on the Akan and the Kingdom of Asante* (Ohio University Press, Athens, 1995)

5

Gold Coast Foodways in the Nineteenth Century

Since the mid-1990s, scholars have debated about how the nineteenth-century transition to legitimate commerce impacted the Gold Coast. Topics of discussion have included the changing nature of slavery, the shifting political landscape, and the economic impact of the commercial transition.[1] It is certain that the coastal Fante (who inhabited the more forested western coast), the Gã (who occupied drier areas in and around Accra), and the Ewe (who settled the eastern coast) experienced changes to their economic, political, and social conditions in the nineteenth century as a result of the cessation of the slave trade. However, even as people from dissimilar regions were coming into greater contact with each other in urban centres and in mission schools, local food retained its important role as a marker of traditional social boundaries and of ethnic difference.[2]

[1] See Robin Law, ed., *From Slave Trade to "Legitimate" Commerce: The Commercial Transition in Nineteenth-Century West Africa* (Cambridge: Cambridge University Press, 1995), esp. the chapters by Ray A. Kea, 'Plantations and Labour in the South-East Gold Coast from the Late Eighteenth to the Mid Nineteenth Century', Martin Lynn, 'The West African Palm Oil Trade in the Nineteenth Century and the 'Crisis of Adaptation' and Gareth Austin, 'Between Abolition and Jihad: The Asante Response to the Ending of the Atlantic Slave Trade, 1807–1896'. See also Parker, *Making the Town.*

[2] Ulrike Sill, *Encounters in Quest of Christian Womanhood* (Leiden: Brill, 2010); Akyeampong, *Drink*; Kea, 'Plantations and Labour'.

B. Simpson Miller, *Food and Identity in Nineteenth and Twentieth Century Ghana*,
Food and Identity in a Globalising World,
https://doi.org/10.1007/978-3-030-88403-1_5

For example, the Fante were identified mainly with their consumption of plantain and yams, the Gã with the consumption of maize, and the Ewe with their consumption of cassava and beans.[3]

The participation of coastal peoples in the global commercial trade from the 1790s to the 1840s necessitated the adoption of three different food production strategies: the production of subsistence food, the provisioning of growing urban markets, and the production of food for both European colonies and urban centres.[4] Although conflict in the nineteenth-century Gold Coast forced many to make concessions as to which starchy staple they came to rely upon, these three strategies enabled a continuity of foodways in terms of essential items such as palm oil and smoked fish. Even as commercial agriculture expanded and access to imported foods grew, local foods never lost their "weightiness".

Lévi-Strauss and Mary Douglas characterised food consumption as a grammar, wherein each meal represented a sentence.[5] This chapter aims to investigate how it was that coastal consumption and identity retained its grammar and associated meanings despite the transition away from traditional starchy staples like millet and sorghum. The chapter is divided into two time periods: the early nineteenth-century transition from the slave trade to legitimate commerce, and the mid-nineteenth to the early twentieth-century transition to British colonial rule. An historical comparison of the specific starchy staples of the Fante, the Gã, and the coastal Ewe will be followed by an exploration of tensions created by Christian commensality customs in missionary schools, and how clashes over these customs made clear existing coastal ethnic identities and social boundaries. Using contemporary primary sources, ethnographies, and oral histories, the food history of these three ethnic groups is examined in a chronology

---

[3] Paul Erdmann Isert, *Letters on West Africa and the Slave Trade*, trans. Selena Axelrod Winsnes (Legon, Accra: Sub-Saharan Publishers, 2007); Hans Christian Monrad, *Two Views from Christiansborg Castle*, trans. Selena Axelrod Winsnes (Legon, Accra: African Books Collective, 2009); Henry Meredith, *An Account of the Gold Coast of Africa* (London: Longman, Hurst, Rees, Orme, and Brown, 1812); William Hutton, *A Voyage to Africa* (London: Longman, Hurst, Rees, Orme, and Brown, 1821); Brodie Cruickshank, *Eighteen Years on the Gold Coast of Africa* (London: London, Hurst and Blackett, 1853).

[4] Kea, 'Plantations and Labour', 120–22.

[5] Claude Lévi-Strauss, *Totemism* (London: Merlin Press, 1964), 64–72; Mary Douglas, *Natural Symbols*; Mary Douglas, *Purity and Danger*.

that roughly coincides with the timing of their encounters with European traders and missionaries.

There presently exists no history of Gold Coast food in the nineteenth century, and this chapter is intended to help fill this gap in the field. The work that comes closest to a historical food study of Ghana is Emmanuel Akyeampong's *Drink, Power, and Cultural Change* (1996). In this monograph Akeyampong used the examination of alcohol consumption from 1860 to 1900 as a lens with which to survey the social changes brought about by the transition from the slave trade to the legitimate trade. His examination of the struggle to control working class access to cheap alcohol tells us much about how ordinary people experienced the changeover to a monetized economy. Utilizing Akeyampong's work as a guide, my own food studies approach will provide an additional perspective from which to review the social history of pre-colonial and post-colonial Ghana in the shift to capitalist production.

Jeremy Rich advanced the field of African urban food history with A *Workman is Worthy of His Meat* (2007). This work is one of the only West African food histories to study the provisioning of a single city from 1840 to 1960. A detailed case study of interdependent food relationships between Libreville and the French empire, Rich used an interdisciplinary approach to examine the often-overlooked feeding of colonial cities. This chapter differs from Rich's approach in its local focus. Whereas Rich was looking at the power relationships in how food was being produced and the role of the French empire in shaping that process, this chapter focuses more upon the maintenance of identities *between* and *within* ethnic groups on the coast.

Historical Context

The situation on the Gold Coast in the early nineteenth century was one of upheaval for many. The British banned the slave trade in 1807, causing an adjustment to an economy where people could no longer sell high-value slaves for trade goods. The Asante, who resided further inland in the forested centre and were once accustomed to going through middlemen like the Fante, now sought to control any residual trade with Europeans directly. They troubled the coastal populations with attacks and occupations for the better part of six decades. Military confrontations between the Asante and the Fante contributed to the growth of British influence on the Gold Coast, as the Fante states signed the Bond of 1844 at Fomena-Adansi, allowing the British to usurp judicial authority from local African courts.[6] British influence was extended in 1850, when they acquired Christiansborg Castle and other forts on the eastern Gold Coast from the Danes. The expansion of administration meant that a wage economy developed, especially in areas inhabited by the Gã.[7] As the slave trade contracted in Fante and Gã country, the Ewe further east on the frontier of the Slave Coast—extending from the Volta River east to Lagos in present-day Nigeria—benefitted from their natural environment of lagoons and rivers to smuggle and trade slaves with the Portuguese and Brazilians until this trade petered out in the 1860s.[8]

The following story from the early nineteenth century tells us a bit of the historical context coastal peoples were navigating whilst growing, cooking, and eating food. Following his enstoolment in 1824, the new Asantehene Osei Yaw Akoto resolved to redouble efforts to dominate the coastal kingdoms, despite warnings to the contrary from the oracular deity, Tano:

> The king sent to ask the oracle of Tanno, the chief fetish [deity] of Asante…He was in reply told to wait till Tanno and his warriors had been to the coast to see whether the king should march down or not. A few weeks later Tanno reported his return from the coast and requested the king to have 100 pots of palm-oil poured into the river Tanno…The oil was accordingly poured into the river, when Tanno said, he had been defeated on the coast by Akra fetishes, and sustained a great loss in killed and wounded, so that the oil was required to dress the wounds of his warriors. The king ought, therefore, not to march against the Akras.[9]

The loss of the spiritual battle by the god Tano and his otherworldly army to the gods of the Gã ('Akras') foretold the defeat of the Asante in battle and tells us of the significance and value of palm oil to pay tribute to and to venerate gods. This story also marks the local sense of place and sets the

[6] G. A. Acquah, *The Fantse of Ghana: A History* (Hull: Hull University, 1957).

[7] Gracia Clark, *Onions Are My Husband: Survival and Accumulation by West African Market Women* (Chicago: University of Chicago Press, 1994).

[8] Akyeampong, *Between the Sea and the Lagoon*, 49; Greene, *Gender*, 141.

[9] Carl Christian Reindorf, *History of the Gold Coast and Asante* (Basel: Missionsbuchhandlung, 1895), 200.

scene of the nineteenth century as a time of turmoil that ultimately caused change in terms of which foods people were willing and able to eat.

There was considerable overlap with respect to palm oil, fish, and salt production among the Fante, Gã, and Ewe. However, as discussed in Chap. 1 the different microclimates on the coast meant that each population specialised in different aspects of food production and therefore placed different starchy staples at the centre of their cuisines. The following chart is a general outline of which ethnic groups ate which starchy staples in the nineteenth century and a brief description of how they were prepared (Fig. 5.1).

Food Name	People/Ethnic Group	Image	Typical Ingredients
kenkey/dokono	Gã/Fante		fermented corn dough steamed in corn husks or banana leaves (firm)
gari (farine de manioc)	Various, but mainly Ewe	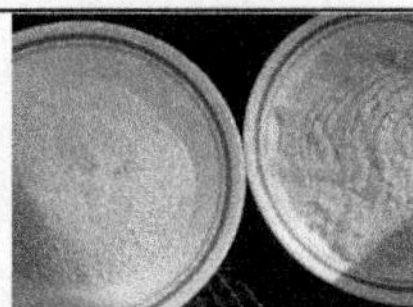	dried, grated, fermented cassava meal
fufu	Fante		peeled, boiled, pounded stiff but elastic dumpling, generally not chewed (yam, cassava, cocoyam, ripe or green plantain, single or combination)

Fig. 5.1 Adapted from "Food and Foodways" in *The Greenwood Encyclopedia of Popular Culture, Osseo-Asare*, 2007. Images from Betumiblog.blogspot.com and the author

The Fante and the Decline of the Slave Trade

By the beginning of the nineteenth century, the Fante were a cosmopolitan people with extensive trade connections in the Atlantic economy. Many could speak one or more European languages, some acknowledged European ancestry, and a few had been educated abroad.[10] Although they continued as prominent merchants and traders, fishing and agricultural production remained important pillars of the economic landscape. Many women made an independent income marketing fish, farming, trading, weaving, pottery, performing rituals and/or healing.[11] Their common food in the eighteenth century was millet, boiled to the consistency of bread or boiled yams, served with a soup of palm oil, herbs, and fish.[12] Soup ingredients in the nineteenth century remained very similar to soup ingredients from two centuries before; fish or poultry, pepper, palm oil, and shallots were accompanied by a "pudding" or fufu, composed of yam or plantain.[13] Kenkey became a staple and prepared market food as early as 1600.[14] Baking of kenkey bread was done initially in the ashes and embers of a fire. Eventually, an earthen or clay oven was used in which a load of wood would be burnt. The preheated oven would be swept clean of its embers and ashes and loaves cooked while the oven cooled.[15] By the early-1800s, both kenkey made of maize, and fufu made of plantain and yams were the starchy foundations of Fante cuisine.[16] Their ecology enabled the Fante to grow a variety of staple starches, thus broadening their selection in comparison with their coastal neighbours.

At the onset of the nineteenth century, the Fante were faced with the prospect of having to change from being consumers to increasing their production of agricultural trade goods like palm oil. The slave trade had been banned, but there was some continuity with respect to the Fante reputation for their involvement in trade. Gold dust, gum copal, and

[10] McCarthy, *Social Change*, 33.

[11] Cruickshank, *Eighteen Years*, 269–81.

[12] W. Bosman, *A New and Accurate Description of the Coast of Guinea* (London: Cass & Co., 1967), 124.

[13] Meredith, *Account*, 88, 111.

[14] De Marees, *Description*, 272.

[15] De Marees, *Description*, 112.

[16] Ibid., 111; Cruickshank, *Eighteen Years*, 274.

ivory from the interior, and more-locally produced palm oil, ground nuts, and malagueta pepper were exchanged for silks, velvets, woollens, earthenware, cutlery, salt, house furnishings, tea, sugar, beer, and spirits.[17] There was no evidence to suggest that items like tea, sugar, and spirits became part of the structure of the proper meal during this period. More than likely, these trade goods were used as ways to display wealth at communal celebrations, to entertain, and in the case of spirits, to facilitate ritual.[18]

Irrespective of the fact that the Fante also consumed kenkey, the very name of kenkey bread indicated its status as a Gã staple, recalling the way in which food was traditionally eaten—a piece of starch was pinched off, dipped in soup, and swallowed, not chewed. The word *kummy*, in Gã-Dangme, a compound of *kɔ*, "bite" with *mi*, "swallow", suggested Gã appreciation of this food as foundational to their cuisine.[19] Kenkey was prepared by wetting maize, grinding it, and allowing it to ferment before being shaped into small loaves, which were then baked, boiled, or steamed.[20] As already stated, both the Gã and Fante consumed kenkey—the former dough including salt and made of balls wrapped in corn husks before steaming, the latter prepared without salt and wrapped in plantain leaves.[21] (The Fante took their rolled-up packets and then baked them "in their small earthen conical ovens".[22]) *Aflatazation* refers to the process of making a portion of the fermented dough into a slurry and cooking it with constant stirring into a thick paste called *aflata*. *Aflata*, then added to the rest of the dough, is what gave kenkey its desirable elastic, sticky texture that particularly suited the traditional method of eating.[23] Kenkey was recognised not only as a fast, convenience food that was satisfying, but its fermentation endowed it with health benefits that made nutrients more readily available and reduced incidence of food-related illness.[24]

---

[17] Ibid., 278–80.

[18] Akyeampong, *Drink*, 87.

[19] Isert, *Letters*, 167; La Fleur, *Fusion Foodways*, 125.

[20] Hutton, *Voyage*, 101–02.

[21] La Fleur, *Fusion Foodways*, 124; 'BetumiBlog: Ghana-Style Kenkey', accessed 18 May 2020, https://www.betumi.com/2007/03/ghana-style-kenkey-italy-has-polenta.html.

[22] Cruickshank, *Eighteen Years*, 275.

[23] Mary Obodai et al., 'Kenkey Production, Vending, and Consumption Practices in Ghana', *Food Chain* 4, no. 3 (October 2014): 275–88.

[24] Patience Mensah et al., 'Antimicrobial Effect of Fermented Ghanaian Maize Dough', *Journal of Applied Bacteriology* 70, no. 3 (1991): 203–10.

Grinding, fermentation, and aflatazation were all very labour intensive, making kenkey production complex and lengthy. One observer of native cooking noted the effort that goes into its preparation: "'Kankie' takes the place of bread, and is made from the flour of native corn, undergoing many operations before being boiled or roasted in plantain leaves and fit to eat."[25] The laborious kenkey processing meant that many were amenable to purchasing kenkey from vendors, who customarily were older women. Indeed, families of women were engaged in the production of kenkey for generations and, as they hired outside help, contributed to furnishing work for the community.[26]

> In the morning, the Negroes usually drink a kind of gruel made of maize meal. This drink is sold and hawked in the early morning hours in public places in the towns, where, moreover, the Negresses sell maize bread, fried fish, all kinds of fruit, soap, baked products or oil cakes, fried beans, salt, etc. These are sold all day, but mostly in the forenoon.[27]

Monrad's account established the pervasive nature of maize vending at the beginning of the nineteenth century. The laborious nature of maize processing contributed to an expansion of cooked-food operations into areas of the settlement outside of the market. This expansion represented local acceptance of the outsourcing of food preparation and a growing demand for readymade foods in urban areas on the coast.

Cassava: A Contested Good

Cassava was introduced to the Gold Coast in the seventeenth century and flourished in the arid climate and infertile soils of the Osu environs of the Danish Fort, Christiansborg. Local experimenters began trying to adapt the introduced varieties of cassava they found near Christiansborg towards types that had a lower poison content (as recognized by taste),

---

[25] George Macdonald, *The Gold Coast, Past and Present* (London: Longmans, Green, 1898), 204.

[26] Obodai et al., 'Kenkey Production'.

[27] Monrad, *Two Views*, 184.

while maintaining attractive features, such as hardiness.[28] Famines caused by war or lack of rain complicated the ability of coastal residents to provide themselves with their customary starchy staples such as maize in the early nineteenth century. The Asante invasion of the Fante in 1806–1807 directly caused the 1809 famine, as people were unable to properly attend to cultivation for the persistent fear of attack. As Reindorf wrote:

> Famines that are still the recollection of old people are those in the year 1809, which was brought about by the Asante invasion of Fante in 1807. Those in 1816, 1822, 1825, 1829, 1832—all started as a consequence of war, at which times people could not properly attend to cultivation, or sometimes by insufficient rain. It was during those famines that many a Fante was sold for a few pounds' weight of corn.[29]

The 1816 famine alone was responsible for the deaths of many thousands of Fante.[30] Fante farmers turned to cassava to buttress themselves against famines by planting the crop in soils that were dry, nutrient-poor, or otherwise unsuitable for maize (which did not tolerate saline spray) or plantains (which required shelter from wind, and moisture).[31] Whatever the Fante thought of cassava before the tumult of the transition to legitimate trade, farmers of the central coast expanded cassava cultivation just after the Asante invasion in 1806–1807 and the ensuing military occupation, which lasted until 1826. Many farmers sought refuge during the first invasion by hiding in the forests, caves, and other agriculturally marginal areas.[32] Cassava, which stored well underground, could be grown in poor soil and retrieved under these arduous circumstances to stave off hunger.

In the lead up to the historic 1826 battle of Katamanso, Asantehene Osei Yaw Akoto sought the assistance of an oracle and the advice of his vassal Boaten (the Omanhene of Dwaben) in the coming conflict with the rebellious coastal Accras. Dwaben was a tributary state located just

[28] La Fleur, *Fusion Foodways*, 163.

[29] Reindorf, *History*, 276.

[30] Cruickshank, *Eighteen Years*, 118.

[31] La Fleur, *Fusion Foodways*, 168.

[32] Monrad, *Two Views*, 102.

southwest of Kumasi, the Asante capital. Notably, King Boaten chose a dish of cassava to represent the coastal enemies of the Asante when consulting the oracle:

> Boaten also sent to consult the oracle of Odente, the highest fetish at Karakye. A hot mess of cassada (or roasted flour) was placed in a dish, with another dish as cover, wrapped up in cloth, and sent to Dwaben with this message: "If the meal is cold in arriving, it means good luck, if warm, Boaten will smell fire on the coast."…The meal was still warm on reaching Dwaben.[33]

Notwithstanding this ill omen, the Asantehene proceeded to make the necessary preparations for war, distributing arms and ammunition to all his warriors, and pouring out libation to the spirits of deceased kings in advance of his march on the coast. The Asante lost the battle of Katamanso, which resulted in the end of Asante suzerainty over many southern and coastal nations. Nevertheless, the use of cassava to represent the Asantehene's enemies was derisively symbolic. Asante historically condemned cassava consumption. Fante proverbs derided cassava as a food even the lowly porcupine was ashamed to eat; *Okuafo afɛre ɔbankye, na kɔtɔkɔnso afɛre*: "The farmer is ashamed to eat cassava and the porcupine is also ashamed to eat it."[34] But because the porcupine was a symbol of the Asante people, the proverb had a double meaning: the Asante considered cassava to be suitable only for the poor, the unfortunate, foreigners, and especially slaves.[35] The Asante's representative use of cassava in seeking prophesy meant that the Fante, were cast beyond the pale of kinship and consideration, placing them at the bottom of the regional social hierarchy.

However much cassava came to be consumed on the coast by the Fante it was never to attain the positive associations that the eating of yam or maize embodied in daily life or in ritual. After the 1803 Danish ban on trans-Atlantic slave trafficking, Danish traders relocated toward the

---

[33] Reindorf, *History*, 201.

[34] J. G. Christaller, *Collection of Three Thousand Six Hundred Tshi Proverbs* (Lewiston: Edwin Mellen Press Ltd, 1990), 244.

[35] La Fleur, *Fusion Foodways*, 168.

Legon Hills and along the Akuapem Ridge near Accra.[36] Associations of cassava with unfortunates in the late eighteenth century are evident in Danish use of slave labour on plantations of indigo, cotton, or sugar cane. Cassava was chosen as a staple on slave provisioning plots and on core plantation grounds for the sustenance of the workers.[37] Over forty years later, cassava was still being used as nourishment for slaves. Scottish traveller John Duncan noted the specific use cassava was put to in his 1847 travel journal: "Slaves in barracoons for exportation are fed twice during the twenty-four hours, morning and evening...They are chiefly fed upon farina, a meal ground from the cassada [cassava] or manioc-root."[38] This account resolutely identified cassava as the chief sustenance for slaves, and reinforced its association with misfortune.

During this difficult time of conflict and famine, the Gã near Accra also turned to cassava to stave off hunger. Gã women and children were seen in Accra foraging for palm fruit and berries. Some had relatives in Akuapem, who provided them with unripe plantains that they smoked and ground for sustenance. Locusts appeared three times in the mid-nineteenth century: in 1833, 1838, and in 1842, exacerbating the already dire conditions.[39] The resulting famines transformed the way farmers planted cassava, especially in James Town and Dutch Town, where large cassava plantations were better able to withstand the assault of locusts. Big local merchants in and around Accra, like the Bannermans, Ankras, and Richters, also increased their plantings of cassava in large gardens close to the towns as a hedge against famine.[40]

Not only was cassava relied upon for subsistence, but it became a way to provision growing urban markets and to absorb slave labour. As

---

[36] C. D. Adams, 'Activities of Danish Botanists in Guinea 1783–1850', *Transactions of the Historical Society of Ghana* 3, no. 1 (1957): 30–46.

[37] Kea, 'Plantations and Labour', 137; Henrik Jeppesen, *Danske plantageanlæg på Guldkystem, 1788–1850* (Place of publication and publisher not identified, 1966), 57–59.

[38] John Duncan, *Travels in Western Africa, in 1845 & 1846* (London: R. Bentley, 1847), 104. Although this account is from coastal present-day Benin, the kingdom of Whydah in the seventeenth century was conquered by the Akwamu, one of the Akan people. It stands to reason that attitudes along the coast toward cassava and its uses would have travelled and been widely known from Accra to Whydah, a distance of less than 200 miles.

[39] Reindorf, *History*, 278; La Fleur, *Fusion Foodways*, 142–44.

[40] Reindorf, *History*, 279.

Europeans in the early nineteenth century began to ban the slave trade, merchants in Accra adjusted to the downturn in the maize trade by using their slaves to produce cassava for the coastal towns.[41] By the end of the nineteenth century, cassava *gari* had become a staple food.

In the late eighteenth century, as maize was the preferred staple, few ways of preparing cassava were developed; according to Isert, "[t]he Blacks prefer to eat these cassavas after they have been roasted in a fire, like chestnuts."[42] Locals attempted to adapt cassava to pre-existing processing and consumption patterns; as with the development of cuisines generally, "cooks picked up ingredients, tools, or techniques that could be incorporated without violating their culinary philosophy."[43] That is to say, initially, cooking techniques for cassava were quite conservative and would follow the processing for more familiar starches like maize or millet.

Initially, many found indigenous cassava to have an undesirable taste and texture when it was processed into the customary *fufu* (pounded like yam) or *ampesi* (boiled). Not least due to its taste and texture, lengthy processing was necessary to diminish its toxicity:

> The transformation of this poisonous plant into an excellent food plant is brought about in the following manner. The root…is peeled and then grated as one grates a horse radish, using a large copper grater. Then it is placed in a cloth and compressed in a press or weighted down with a large stone until not more moisture comes out. The pressed residue is then thrown into a pot and dried over a fire, a process which cooks it a little. [*garri*/*farin*] It is then ready for use. Others have an iron plate for this purpose, on which they usually bake the pressed root in thin cakes as we bake waffles.[44]

These processing methods were attributed to Afro-Brazilians who resettled in West Africa following a series of slave revolts in Bahia between

---

[41] Kea, 'Plantations and Labour', 125.

[42] Isert, *Letters*, 168.

[43] Rachel Laudan, *Cuisine and Empire: Cooking in World History* (Berkeley: University of California Press, 2015), 5.

[44] Isert, *Letters*, 246.

1831 and 1835.[45] The dish now known as *gari fortor*,—most likely derived from the Brazilian-Portuguese *farofa*: grated roasted manioc—which mixes flavourings like onions, tomatoes, and eggs into the shredded cassava before frying, has become a residual marker of the Brazilian contribution to today's Ghanaian cuisine.[46]

The coastal Ewe did not have the same reservations about the consumption of cassava as the Fante did. As stated in Chap. 1, the sandy soil of their coastal environment meant that cassava became the main starchy staple of the Ewe, with maize being an important second starch. Keta (literally meaning "on top of sand"[47]) became a major fishing town and surf port, a link for trade with the Slave Coast to the east. The people who resided in the area east of the Volta were different from their coastal neighbours, like the Fante, in that production was always a primary concern for them. Many who migrated to the lagoons of the eastern coast in the mid-seventeenth century had to adjust from a rural land-bound lifestyle to an area of tide waters and lagoons.[48] People learned to specialise in fishing and salt-making while absorbing the culture and religious beliefs of the indigenous coastal peoples. Having adopted new processing methods for cassava, Ewe women showed innovation with respect to the mechanisation of the removal of toxins. Using heavy weights instead of manual labour, the longer mechanical processing produced a sweet-sour taste caused by fermentation, that locals preferred.[49]

Along with the adoption of cassava as a staple starch, the Ewe embraced legumes as a principal food. Nineteenth-century sources frequently mention Ewe production of legumes for their own subsistence: "There are various kinds of pulse, such as lacros, which is brown and not unlike

[45] Paul E. Lovejoy, 'Background to Rebellion: The Origins of Muslim Slaves in Bahia', *Slavery & Abolition* 15, no. 2 (1 August 1994): 151–80; Lisa A. Lindsay, '"To Return to the Bosom of Their Fatherland": Brazilian Immigrants in Nineteenth-century Lagos', *Slavery & Abolition* 15, no. 1 (April 1994): 22–50.

[46] Fran Osseo-Asare and Barbara Baëta, *The Ghana Cookbook* (New York: Hippocrene Books, 2015), 152.

[47] Akyeampong, *Between the Sea and the Lagoon*, 28.

[48] Ibid., 31.

[49] E. V. Doku, *Cassava in Ghana* (Accra: Ghana Universities Press, 1969), 33; La Fleur, *Fusion Foodways*, 167.

black beans. A kind of gruel is made of these."[50] Monrad described the Ewe diet of those living near the Volta River as based primarily on beans and included other protein sources such as fish and turtles.[51] The Ewe found consumption of beans and cassava as the best way to ensure food security in the nineteenth century, and these foods became part of the sense of place they built within the strictures of the coastal environment they inhabited.

In contrast to Rich's study of Gabon in the nineteenth century, the issues on the Gold Coast did not revolve so much around who was producing the food and how it was being transported, but it was more about who was eating which foods and why in the context of opposition to other ethnic groups, both slave and free. The nineteenth century was a time of tough choices and changes in terms of what people were having to cook and eat. Anyone that could afford to did not consume cassava and, despite its increased and varied consumption, it was never elevated to the status of ritual use the way other starches were.[52]

Legitimate Trade

With the late nineteenth-century transition away from palm oil exports to gold and cocoa, rice and wheat bread moved into the frame to become staple foods on the coast. By the time Carl Reindorf published his *History of the Gold Coast and Asante* in 1895, wheat bread had become a staple by which he measured the rising costs of provisions in commercial and market towns.[53] Imported rice also came to be considered an essential starch, which was brought in by the ton from other British colonies like Sierra Leone and India.[54] Imported rice, wheat flour, tinned meats, and tinned tomatoes—and the subsequent efforts by the new colonial government to

[50] Monrad, *Two Views*, 140.

[51] Monrad, 139–40, 160.

[52] La Fleur, *Fusion Foodways*, 179.

[53] Reindorf, *History*, 279.

[54] Earl of Kimberley to Colonel Harley,11 July 1873, CO 879/4/493, TNA.

decrease dependence upon these items—mark them as noteworthy phenomena in the history of coastal cuisine.

For the Fante on the central Gold Coast, their ongoing conflict with the Asante contributed to the adoption of imported staple starches into their repertoire. The changeover from the slave trade to legitimate trade produced different relations between the Fante and the Europeans. In 1844, the Fante signed The Bond with the British whereby they agreed that all murder, robbery, and other serious crimes be tried by British judicial officers.[55] This affected their ability to collect money from their own native courts and to recirculate luxury goods, as they had done in the past during the slave trade. In addition, by 1851, the shorter stay of European merchants did not leave much opportunity for chiefs and merchants to forge relationships. Following the 1868 exchange of territory between the British and the Dutch, the latter finally departed from the Gold Coast in 1872, leaving the British as the sole European power. The erosion of the slave trade and the Fante's place as middlemen continued to decline over the century, and this would come to be reflected in their foodways.

Continuing conflict on the coast in the early 1870s necessitated the importation of rice in the support of Fante soldiers fighting the Asante. After the Fante experienced a military defeat on 6 June 1873, they were sent 90 tonnes of rice, 18 tonnes of biscuits, and nine tonnes of preserved meats for "hospital use", intended for injured soldiers. In addition, the Fante's British allies deployed steam ships to carry relief provisions to the 30,000 Fante refugees driven toward Cape Coast seeking shelter from Asante attack. Fante troops stationed at Elmina were supplied with rice enough for twenty men. Stores were divided up into "food for natives" and "food for Europeans". Native food consisted of palm oil, kola nut, maize, rice, and wheat flour.[56] Thus, although locals continued to consume local foods like palm oil and maize, rice and wheat flour edged into the repertoire and began to be viewed as necessities.

Steam shipments of imported supplies continued for several years, rice being among the most highly prized items. The following account revealed that rice was valuable enough to attract the interest of thieves:

---

[55] Acquah, *The Fantse of Ghana*, 54.

[56] Earl of Kimberley to Colonel Harley, 11 July 1873, CO 879/4/493, TNA.

> Just inside the doorway were three women reclining upon some large bundles. The police assisted these ladies to rise and behold they had been sitting upon bags of Navy biscuit…we discovered a row of small sacks leaning against the walls. A rush was made at these, but, to the disappointment of the police, they were found to contain only Indian corn. The head man now became more virulent than ever in his abuse, when one man made a rush at the sacks, and, plunging his arm into one up to the elbow, produced a handful of Government rice.[57]

The headman of the raided house was a Euro-African clerk who intended to capitalise upon the high demand for rice. Therefore, British relief efforts stimulated demand for rice, and the value of rice and wheat flour continued to rise. Rice, once harvested and processed, was fast, easy, and simple to prepare, requiring only boiling water. A preferred complement to the dish of "palm oil chop", rice was often substituted for local starches on the Fante coast, further encouraging its consumption.[58]

The growing reliance upon imported starches coexisted with the consumption of local starchy staples. The Annual Colonial Report for the Gold Coast in 1895 disclosed that imports of provisions like rice and flour had more than doubled since 1890. Imports continued to rise, notwithstanding the 10 percent ad valorem tax on imported foods and other luxuries. However much rice and flour consumption increased, the report also demonstrated that people continued to rely upon maize, yams, cassava, and plantain as staple foods, and that internal trade in palm oil remained strong.[59] These figures suggested that imported rice and wheat had not completely supplanted local staples.

Irrespective of the dual use of both imported and local staples, the commercial trade and concurrent growth of urban centres contributed to a growing concern over the rising costs of food and a fear over the neglect of the production of local foodstuffs:

---

[57] A. B. Ellis, *West African Sketches* (London: Tinsley, 1881), 56.

[58] Richard Francis Burton, *Wanderings in West Africa from Liverpool to Fernando Po* (London: Tinsley Brothers, 1863), 146.

[59] F.M. Hodgson, Gold Coast Annual Report for 1895 (London: Parliament, February 1897).

> The Krobos are known to be the best and able farmers on the Gold Coast; all the inhabitants of about 40,000 in number are engaged in farming, viz., palm-oil making, which obliges them to have all their lands planted with palm-trees. Although they buy thousands of acres of land from the Akuapems, Akwamus, and Akems, but these lands are so distant from towns that provisions there obtained can scarcely be conveyed to market, hence provisions are dearer in Krobo than even on the coast.[60]

Anxiety over the cost of food in towns and the fear of dependence on imports in areas where the commercial trade was expanding were justified. Over the next decade and a half, the colonial government continued to wrestle with the growing popularity of rice consumption, which led to the juggling of rice imports, at times favouring rice from Sierra Leone over Indian rice, depending upon which was cheaper to import.[61] Toward the end of the nineteenth-century commercial transition, the price of staples such as rice, wheat flour, and even local starchy staple foods like yam fluctuated depending upon the prices of exports such as palm oil and cocoa. From 1891 to 1931, cassava was the only food tracked by the Gold Coast census that grew cheaper over time, indicating that commercial trade contributed to staple food price volatility in the colony.[62]

Even though consumption of cassava, rice, and wheat grew steadily, none of these foods were considered appropriate for use in rituals to honour gods and/or ancestors. Indeed, local starches retained their place at the centre of the main meal of the day. Cruickshank had this to say about the foods the Fante spent time and energy cultivating: "More care is taken with their yam and cassada [sic] plantations, which are carefully kept clear of weeds, and staked and trimmed."[63] Even though cassava was part of the food strategy employed by the Fante,

---

[60] Reindorf, *History*, 280.

[61] Governor Rodger to Earl of Crewe, Memorandum on Sierra Leone Rice, 16 December 1909, ADM 11/1/89, PRAAD Accra.

[62] Jonathan E. Robins, '"Food Comes First": The Development of Colonial Nutritional Policy in Ghana, 1900–1950', *Global Food History* 4, no. 2 (July 3, 2018): 168–88.

[63] Cruickshank, *Eighteen Years*, 274.

earlier contemporary sources observed the Akan "yam festival" or Odwira remained at the centre of Fante ritual practice. The Fante from Dutch and Danish Accra also came to English Accra during *Homowo* to feast on yams once they reached full maturity.[64] Stools were shrines of the ancestors that were kept in sacred rooms and, in the case of the Fante, were fed with *eto*—mashed plantain and yam. Foods that people cooked and ate were entities that drew their significance from the wider contexts in which they appeared, and the contexts these new foods appeared in—the food of unfortunates or that of foreigners—were not considered fitting for ritual use.[65]

Yam was eaten on the coast in ways that honoured the social and divine sources and memberships of life, and in ways that reflected its multiple histories.[66] The Gã also used yam to honour stools and war gods brought from neighbouring Akan regions, that were meant to protect them from the aggression of their neighbours: "Amid song, dance, gambling and carousing is held the sacrifice to the fetish of yams, maize, and gobbegobber (a variety of red beans) [peanuts], etc., and as long as the festivities last, yams are eaten, as well as other products cultivated in the earth at that time of year."[67] Relatives from the bush farms would bring presents of yam which were laid outside the stool room. *Oto* [or *eto*], made of yams and egg, was placed on every stool.[68] It was important that these foreign entities from the forested centre were fed with the foods which were tied to their land of origin, the only foods considered acceptable for their sustenance.

There were exceptions to the rule of using traditional starches in ritual. For example, irrespective of maize's relatively "new" status, women were seen propitiating the gods of the sea with "heave-offerings or corn-bread, mixed with palm oil [*kpeikpei*], upon all the rocks and principal

[64] Meredith, *Account*, 30; Hutton, *Voyage*, 325.

[65] Norman Wirzba, *Food and Faith: A Theology of Eating* (Cambridge: Cambridge University Press, 2011), 5.

[66] Ibid., 13, 17.

[67] Monrad, *Two Views*, 56.

[68] Margaret J Field, *Social Organization of the Gã People* (Accra: Crown Agents for the Colonies, 1940), 179.

Fetishes."[69] Much like the adoption of cowpeas by the coastal Ewe as a ritual food, maize too played a major part in the daily sustenance and ritual life of the Gã in the nineteenth century. Through growing, cooking and eating these foods over time, people on the coast developed local customs and moved to an experience of terroir—and experience of the unique flavours of the microclimate—and imagined a world with these foods. These imaginings were eaten, spoken, shared, and adopted into ritual life.[70] The growing, cooking, and eating of cassava however, situated and defined people as unfortunates for the Fante and Gã, who had long made a living off the slave trade and had other options for starchy staples.

Palm oil emerged by the 1820s as the principal 'legitimate' commodity that replaced slave exports. The Fante, Gã, and Ewe all accepted the gift of palm oil as an integral part of their foodways. It was historically considered to have medicinal properties as a salve and in keeping stomachs and intestines in healthy condition.[71] Despite the commercial expansion of the palm oil trade, the use of palm oil provided continuity in the cooking and eating habits of all three ethnic groups on the coast throughout the nineteenth century. There are a variety of types of palm oil, and processing methods produced differences in taste and quality. There was the palm kernel oil of the inner seed often used in soaps, as well as the pericarp or pulp of the palm fruit. "The fruit has a colour like paprika or glowing coals, with the softness of red velvet, the silkiness of a fine sari, and the richness of fresh cream."[72] The creamy palm pulp, or butter, was considered unparalleled and used in the preparation of soups, while the oil was reserved for frying and making stews (Fig. 5.2).

---

[69] Cruickshank, *Eighteen Years*, 188–89.

[70] Wirzba, *Food and Faith*, 190.

[71] Johannes Rask, *Two Views from Christiansborg Castle*, trans. Selena Axelrod Winsnes (Legon: African Books Collective, 2009), 56.

[72] Osseo-Asare and Baëta, *Ghana Cookbook*, 108.

Fig. 5.2 Frying shitto in palm oil in Accra

Palm Oil "Chop"

Oil Palm Nuts
Boiling water
Meat
Sieve

- Two or three handfuls of the ripe nuts of the oil palm are cooked in a sufficient amount of water until their fibrous, pithy, oily part is very soft and most of it has come apart in the water.
- Pour the entire affair into a hair sieve and through this strain everything that has loosened from the stones and fibres.
- Cook the meat of chickens, sheep, goats, monkeys, or other game in the sieved sauce, and when it is finished pour this vegetable-meat soup over the plantain dumplings (*foifoi*).
- Then the dish is ready, except that it needs the further addition of salt and Spanish pepper.

Adapted from Paul Erdmann Isert, *Letters on West Africa and the Slave Trade*, p. 290.

As noted in Chap. 1, palm oil was considered essential to the preparation of most dishes on the coast. In the nineteenth century, Meredith observed that pepper was "a universal and necessary ingredient" to fish or poultry soup "to which are added palm oil recently express [sic], pepper, salt, and eschallots."[73] Indeed, to this day, the palm oil and pepper condiment *shito/shitto* is still considered a vital addition to meals and is ubiquitous on tables in restaurants today.[74] Palm oil was mixed into breads and cakes, used to fry fish and other meats, and featured in palm nut soup, a foundational dish for the Fante, Gã, and Ewe.[75] A common dish in Accra was fish fried in palm oil. Contemporary sources throughout the nineteenth century also stated that smoked and salted fish were consumed in palm nut soup.[76] Palm oil was locally prized for its ability to satisfy, and it provided a pleasing richness and fullness of texture to soups and stews. Palm oil then was an item that historically characterised coastal cuisine.

Smallholders leveraged durable institutions to expand their northern trade prospects of palm oil, fish, and salt during the transition to the legitimate trade. In so doing, they aided in the preservation of local foodways. In the early nineteenth century, palm oil came mostly from Krobo and Akuapem in the eastern part of the country.[77] Existing political, economic, and social structures facilitated smallholder control over the movement of goods—in this case palm oil—to market, both for internal consumption and for global export. Like the production of cassava, palm oil production aided in the absorption of the surplus slave labour available during the long decline of the slave trade. Additionally, it continued to be produced for local consumption at the same time demand for export to Europe was being met by smallholders in the hinterland. "Rather, it was precisely because the oil trade represented continuity rather than change and because its structures and practices were a continuation of the old, that it grew so rapidly in the early nineteenth century."[78] This ability to meet demand was largely due to its foundation in the slave trade provisioning system that preceded it, exploiting the same trading organisations, techniques, and ports.

---

[73] Meredith, *Account*, 111.

[74] Osseo-Asare and Baëta, *Ghana Cookbook*, 236.

[75] Monrad, *Two Views*, 158.

[76] Ibid., 175; Meredith, *Account*, 88; Hutton, *Voyage*, 92; Cruickshank, *Eighteen Years*, 275.

[77] Meredith, *Account*, 221n.

[78] Lynn, 'West African Palm Oil Trade', 60–61.

Palm Nut Soup—*Nkawan* (Gã)

Dried Stink Fish
Palm Nuts
Boiling water
Salt & Peppers
Onions
Vegetables

- Boil palm nuts till soft and pound in wooden mortar. Boil again until palm oil and palm liquid separate from the water.
- Into this palm oil and palm liquid put dried fish, salt, peppers, onions & other vegetables and boil until vegetables are soft.
- Vegetables can be fished out, pounded, reintroduced and the soup boiled again.

Adapted from M. J. Field, 'Gold Coast Food,' 18.

The marketing of palm oil provided a living for many along the coast, including the Fante, who began migrating to urban centres in the mid-nineteenth century due to prospects resulting from the palm oil trade.[79] The Gã also profited from the trade of palm oil through Accra, and through their expansion of the trade to Prampram and Ada near the mouth of the Volta.[80] Although the Fante were known as commercial palm oil traders near the coast, the Adangme-speaking Krobo (roughly 80 miles northwest of Ada on the Volta River) became leading producers and exporters of palm oil in the nineteenth century. Refugees of the slave trade who migrated to the Krobo Mountain in the seventeenth century, they consumed traditional millet, cow peas, dried fish, game, and palm oil. Palm oil was traded commercially to their neighbours possessed of less fertile territory, closer to the coast.[81] After the defeat of the Asante at Katamanso in 1826 (ten miles southwest of the mountain), the Krobo felt safe to venture further afield into the Akuapem hills.[82] The effective use

[79] Cruickshank, *Eighteen Years*, 59.

[80] Parker, *Making the Town*, 57–62.

[81] Louis E. Wilson, 'The "Bloodless Conquest" in Southeastern Ghana: The Huza and Territorial Expansion of the Krobo in the 19th Century', *International Journal of African Historical Studies* 23, no. 2 (1990): 97.

[82] Major J. J. Crooks, *Records Relating to the Gold Coast Settlements from 1750 to 1874* (Milton Park: Routledge, 2013), 230–50.

and expansion of their small family parcels (*huza*) into Akuapem enabled them to become the leading commercial exporters of valuable palm oil, production of which doubled from the period between 1826 to 1841.[83]

Palm kernel oil (expressed from the hard-inner kernel) also brought profit as it was processed and sold exclusively by women. Both second (or white) palm kernel oil and brown (or black) palm kernel oil was used domestically for frying and making stews, and for sale in local markets, as well as for commercial export. It was primarily produced by the labour of women, and their children and slaves, who worked production into other daily chores.[84] Domestic palm oil processing was principally women's work, with men keeping the commercial profit from the palm fruit.[85] Women directly reaped the benefits of the establishment of palm oil commercialisation as they often managed to earn enough to send their children to mission schools, gaining them places in the future administration of the colony.[86]

Traditional structures and domestic economies contributed to the success, survival, and expansion of palm-oil production for both men and women: "small-scale trade was not only an engine for national economic growth; it provided an indispensable preparation for large-scale entrepreneurship. The net effect of all these changes was a widened opportunity structure for young men anxious for careers in commerce at principal towns up and down the…coast."[87] Women who provisioned the palm oil business were part of this expanding prosperity. The central role of palm oil in coastal cuisine and the expanded opportunities for the marketing of cooked food, combined with resilient structures of trade, credit, and communication facilitated the transition away from the slave trade. Despite the commercial expansion in the export of palm oil, locals met the expanding demand and continued to supply palm oil as a trade item to the interior, as well as meeting their own consumption needs and the

[83] Wilson, 'Bloodless Conquest', 269–97.

[84] Hugo Huber, *The Krobo: Traditional Social and Religious Life of a West African People* (St. Augustin: Anthropos Institute, 1963), 61.

[85] Maier, 'Precolonial Palm Oil Production', 32.

[86] Wilson, 'Bloodless Conquest', 293.

[87] Raymond E. Dumett, 'African Merchants of the Gold Coast, 1860–1905: Dynamics of Indigenous Entrepreneurship', *Comparative Studies in Society and History* 25, no. 4 (1983): 668.

needs of growing urban centres. It remained a central feature of the coastal cuisine in the nineteenth and into the twentieth century.

Another nineteenth-century staple and important source of income for local women was dried fish. Dried fish has been traded from the coast to the forest interior for centuries. The Akuapem were known to trade with the coastal people for dried fish: "Like most of the inhabitants of the coast, the Akras fish in the sea, as well as in the lagoon. They catch great quantities of fish which they consume both fresh and sun-dried. The most common fish of all is the herring…The fish are dried and sold at a good price to the people living further inland."[88] The mountaineers traded powdered cam-wood (red sandalwood) mixed with palm-oil—used for rituals—in return for large quantities of smoked and dried fish.[89]

The fish trade, essential to the foodways of the region, continued throughout the nineteenth century. Monrad noted the locals near Christansborg had a "taste for half-rotten" or, as he called it, "snuf fish".[90] Critically important as a source of protein and as a flavouring for soups and stews, in places such as Winneba and Accra the fishermen had a thriving trade, particularly in the porguis fish, which was processed by drying in the sun: "They spread them over the towns to dry, so that… it is almost difficult to walk."[91] Hutton noted that in addition to its use in making "excellent soup", locals could not consume half of what they produced, transporting considerable quantities to neighbouring regions. Demand for fish grew over time during the expansion of the commercial trade. By the 1840s, growing fish consumption in expanding urban areas necessitated the importation of fish: "In addition to what they cure for themselves, there is a very considerable supply imported by the Americans, which is eagerly purchased."[92] Expanding commercial trade and growing urban centres stimulated the importation of fish, much as it had done for foundational starches.

The growth in demand for fish intensified once colonial rule took firmer shape in 1874 when, following the departure of the Dutch and the

---

[88] Isert, *Letters*, 168.

[89] Reindorf, *History*, 131.

[90] Monrad, *Two Views*, 112.

[91] Hutton, *Voyage*, 92.

[92] Cruickshank, *Eighteen Years*, 275.

military defeat of Asante in the war of 1873–1874, the British declared a formal Crown Colony on the Gold Coast in 1874. The transfer of the colonial capital to Accra in 1877 soon followed. In the new Crown Colony, colonial capitalism continued to expand, a process which further accelerated with the emergence of cocoa as a dynamic new export crop from the 1890s. Colonial enterprises developed a wage economy, attracting workers to places like Accra which, in turn, drove demand for provisions like fish.

In the twentieth century, the long tradition of fish trading on the coast was to encounter a colonial obstacle. In 1909, the Accra Town Council decided to intervene in the marketing of stink fish by adopting a policy of destroying the fish ovens used to smoke it. In a coordinated protest, a letter from Koley Ambah, dated 27 February 1909 stated that on the 14 November 1908 "officers of the Accra Town Council took the liberty of going from house to house and breaking all our ovens in which we roast and smoke our herrings bought from fishermen for trade purposes".[93] Madam Ambah argued that the smoking of fish was necessary for the preservation of this vital protein source, and that the Council taxes were paid out of the trade of such foodstuffs. Furthermore, she reminded them that merchants, administrators, and labourers of the Accra Railway Co. "buy from sellers and eat immediately they break off from business", implying that the destruction of the ovens had far-reaching provisioning and economic consequences for the local economy. This letter was accompanied by a petition signed by 59 other women. The Gã Manste, Tackie Obiri, also sent a letter, dated 3 August 1909, to negotiate with the council as to the construction of new ovens on the outskirts of town. He petitioned the Council that Health Officers under John Maxwell's administration gave people leave to "carry on the industry as it carried no disease", reminding them that the smoking of fish posed no danger to the public health of the town and that it was vital to the local and regional economy.[94]

In connection with the town council sanitation actions, Maxwell received a visit from a group of women who marketed smoked fish: "Palaver with boys re Biscuits…Amusing scene, 2 women called re

[93] Koley Ambah and Ga Manste Tackie Obiri, Fish Ovens—Destruction of by Accra Town Council, February 1909, ADM 11/1/78, PRAAD Accra.

[94] Ambah and Obiri, ADM 11/1/78.

demolition of their fish ovens, whilst I was having tea. Women looked fierce, but gave them all biscuits and promised to look into their complaint and they all went away happy."[95] Maxwell's diary entry called attention to the persistent efforts of the new colonial government to regulate food processing spaces in urban Accra and the sustained resistance the locals presented to their efforts. Resistance to the destruction of the fish ovens continued right up to Maxwell's retirement in 1930, a further testament to the central importance of this dried and smoked fish to the local cuisine and to the internal trade.

Not just the processing of fish, but the method of fishing was vitally important to both the Gã and the Fante in maintaining their cultural traditions. Knowledge of the ritual topography and its proper sacramental manipulation was necessary to successfully manage local ecology.[96] Both the Gã and the Fante were known to use cast nets—weighted circular nets thrown from the shore over the catch, and then drawn in snagging the catch inside. They also abstained from fishing on proscribed days of the week in order to avoid overfishing.[97] "The cast-net is commonly used all over the coast, and there are plenty of fish of almost all descriptions. In the month of September, the natives of Winnebah, Accra, and other places, go about twenty miles out to sea, to the Porguis Bank, where they are sure of loading their canoes every day (except their fetish day) with porguis."[98] Spacing of fishing days was closely managed and integrated into the social organization on the coast.

A perusal of conflicts surrounding fishing methods demonstrated the enduring struggle of local leadership to maintain their traditions in the face of the introduction of new technologies. On 23 February 1909, District Commissioner C. H. Hart Davies took the minutes during a meeting over a skirmish between the Fante fishermen of Anomabo and those of Saltpond (Kormantin). At issue was the Anomabo's attack of Saltpond fishermen for using Ali, or drift nets,—a technological innovation introduced after 1890 from Nigeria—instead of conventional cast

[95] Diaries of Sir John Maxwell, 9 July 1909, MSS.Afr.s.2133, Box 1, Bodleian Libraries, Oxford, United Kingdom.; Fish Ovens, ADM 11/1/78.

[96] Parker, *Making the Town*, 21.

[97] Monrad, *Two Views*, 241; Hutton, *Voyage*, 55.

[98] Ibid., 91.

nets. Drift nets hung vertically in the water like a curtain with floats attached to a rope along the top of the net, and weights attached to another rope along the bottom of the net. The two groups had an agreement that Ali nets would not be used outside one's own village waters. The use of the drift nets was a violation of this agreement. The Host Captain of Anomabo stated: "We fear Ali nets will catch all the fish in the sea. We don't 'know' Ali nets, nor want them. Let people use them in their own waters only".[99] The issue was resolved with government mediation whereby seven offenders from Anomabo were made to pay ten pounds each for damage to the Saltpond Ali nets and further fined fifteen pounds each for the violent attack. Any future disagreements were to be referred to Native courts for resolution, bypassing traditional authority structures. Trouble broke out again in 1916 when the colonial government authorised the use of Ali nets. Riots erupted in response to the authorization as locals feared the depletion of fishing stock and the usurpation of traditional spiritual and commercial structures.[100]

The previous incidents succinctly illustrated the underlying spiritual and commercial importance of the catching, processing, and consumption of fish to coastal peoples, which remained a constant factor in the foodways of the nineteenth century. The role of the chief fisherman was a complex one whereby he was responsible for the transmission of the art of fishing, for appeasing the gods, and for the governance of the community. This is true for both fishing in the sea and in the lagoons. He alone decided who could stay in the town, fish in the local waters, and what sorts of nets were to be used.[101] He officiated at the opening of the fishing season in a ceremony whereby he sowed the millet seeds that foretold the prospects for the coming year. Additionally, he appointed each man his proper day and turn, and collected fees for the right to fish in the area under his jurisdiction.[102] The colonial authorisation of the use of Ali nets represented a direct violation of sacred local custom designed to ensure

[99] C. H. Hart Davies, Kormantine Fishermen Attack Made on by Anomabo Fishermen for Using Ali Nets in Their Waters, 16 February 1909, ADM 11/1/80, PRAAD Accra.

[100] 'Supreme Court Fishing Nets Decision Riots', 1916, ADM 11/1/628, PRAAD Accra.

[101] Field, *Social Organization*, 134, 156.

[102] Frederick Robert Irvine, *The Fishes and Fisheries of the Gold Coast* (London: Crown Agents for the Colonies, 1947), 29.

prosperous and fair fishing, processing, trading, and consumption for all. The usurpation of the chief fisherman's authority over this vital food resource—which also happened to be the foundation for the local foodways and economy—was successfully resisted by locals, ensuring the continuity of local foodways and intraregional trade in the nineteenth century. Nineteenth-century market food reports and their recorded opposition to colonial sanitation initiatives conclusively demonstrated that essential ingredients, such as dried fish and palm oil, resolutely remained at the centre of Gold Coast cuisine.

Missionary schools in the nineteenth century were other places where tensions between local foodways and social structures came into conflict with European ideas about what was appropriate.[103] Beginning in the 1830s, Wesleyans at Cape Coast and Basel Missionaries at Akropong established schools as part of their missions.[104] Wesleyan Methodist missionaries began to provide education for girls from 1844, where they were taught the womanly arts of housekeeping.[105] The non-denominational Basel Mission began work on the eastern Gold Coast in 1829, and by 1848 had founded a training college at Akropong in Akuapem to produce teacher-catechists. It remained the only teacher-training college on the coast for more than 50 years and drew students from every region. By 1852, Reverend T. B. Freeman boasted that Adoo, the chief of Mankessim, was to place 15 children from his own household in the Wesleyan mission school, which was established on the ruins of an important shrine there.[106] Mission schools then provided exposure to both foreign and regional foodways and identities.

Missionary schools illuminated conflict over the role of women as primarily responsible for the social reproduction of identity. Whereas there were boarding schools for boys, girls were typically placed in missionaries' households in order to train them closely in precepts of European Christian womanhood. They were called "girls in training". It was not until 1860 that girls' education was institutionalised. In institutional

[103] Cruickshank, *Eighteen Years*, 102–06.

[104] Reindorf, *History*, 225–27.

[105] McCarthy, *Social Change*, 14.

[106] Cruickshank, *Eighteen Years*, 334.

settings, the observation of ethnic food taboos, peer group commensality, and social status expressed through the provisioning, cooking and eating of food increasingly became an issue for the pupils.

As previously discussed in Chap. 2, one's ability to eat a type of food revealed one's family group, clan, or lineage, regional origination, and one's social status. In the late 1840s, food arrangements in the Widmann missionary household in Akropong were the responsibility of mothers, who had food sent to their daughters residing there as girls in training. When Rosina Widmann changed this system to a more commensal approach where students all shared a table with the Widmanns, parents expressed their concern over the students' adherence to clan or lineage food taboos. This was in keeping with their responsibility for providing appropriate food for their children and to ensure that social norms were adhered to. In response to these concerns, the policy was changed in 1850. Students were given *kostgeld* or food money to purchase food directly from vendors.[107] This satisfied mothers' concerns that students retained control over the maintenance of their family and clan identities via their food consumption.

It was with the Basel Girls Institutes in the 1860s that tensions boiled over around the traditional role of women as breadwinners and nourishers of their progeny. The Basel mission made the decision to standardise their schools with the use of textbooks and the adherence to a general syllabus and timetable. As part of the standardisation effort, it was also decided to place girls in boarding schools instead of in the homes of missionaries. In 1860, the Girls Institute at Abokobi, just north of Accra, decided to require commensality—meaning all students were to consume food cooked in the institution all together at the same time. This was part of the strict new timetable designed to minimise lengthy mealtimes in favour of additional time for classroom instruction. The students revolted in reaction to this new commensality requirement. Seven ran away, and enrolment declined precipitously in subsequent years.[108]

As discussed in Chap. 2, one's peer group was defined by one's age and ethnic or geographic origin. Students would customarily eat in similar age

---

[107] Sill, *Encounters*, 296.

[108] Ibid., 287–309.

and ethnic groupings in order to preserve the appropriate social distances. The timetable in 1858 allowed for the midday meal to be from 11am–2pm. The evening meal was scheduled from 4–6pm. This allowed enough time for students to prepare or fetch meals, and time for eating.[109] The subsequent changeover two years later to commensal meals represented a violation of peer groups customs. Although commensality was more efficient, it violated peer group eating norms and prevented the observation of food taboos, which students found extremely objectionable.

Not only was the revocation of peer group eating at issue, but the disregard for local social ideas about dependency was problematic. Local custom dictated that slaves were only to be compensated with food and clothing. The schools' attempt to provide food for students by sourcing it, cooking it, and serving it to them was akin to reducing the students to slave status. This case provides an interesting contrast to Rich's findings about status and worth in Libreville of the 1870s. Due to the isolation of Libreville and lack of adequate transport from the hinterland, residents of Libreville who worked for missions or traders came to consume imported food, which was often easier to obtain (although much more expensive) than local food. The workers came to expect to be provisioned with imported food, which signalled to them their social status according to their level of education, and their value to the employer.[110] By contrast, Gold Coast locals had access to local food, the trade of which underpinned their economy and whose preparation and consumption signalled their social identity and rank. Additionally, it was the mother's role to provide properly cooked food for her progeny. The school was applying pressure on these relational dependencies in trying to provide food for students and requiring commensality. Controversy in new institutions, such as missionary schools, demonstrated that local foodways still underpinned cultural identities. The woman's role as the provider of food for dependents, local commensality customs, and continued observation of food taboos indicated that local ways of eating, and ideas of consumption endured during and after the transition to the legitimate commercial trade.

---

[109] Johann Dieterle, Basel Mission Report Aburi, ABM, D-1,9, PRAAD Koforidua.
[110] Rich, *A Workman Is Worthy of His Meat*, 90.

The historical analysis of commensality in mission schools confirmed that food and eating habits were an important way of transmitting culture and ethnic identity to children.[111] Food habits were often the last traits to be lost in the acculturation process, as eating usually took place in the private domain and among other in-group members.[112] It is where food, meals and associated customs are transformed into 'the essence of identity'.[113] Food habits are often difficult to change as food has important psychological links with place, family, and community. In summary, foods arouse intense emotional responses.[114] The attempts of mission schools to change commensality and provisioning habits ran afoul of local ideas of eating and drinking at the same table in the nineteenth century. Local eating habits which were structured around social groupings of seniority, ethnicity, and gender, proved resilient in the face of new cultural institutions such as mission schools and Christianity.

Conclusion

This chapter identified the ways in which the Fante, Ewe, and Gã adjusted their cooking and eating traditions during the transition to the commercial trade in the nineteenth-century Gold Coast. The choices of staple starches changed, but the composition of the ideal meal and its concurrent signalling of social roles and ethnic identities remained a constant during the period. In addition, important components of the main meal, such as palm oil and dried fish persisted in the foodways. New starches, like American cassava and imported rice and wheat flour, slotted into situations where either the environment did not allow production of traditional starches, or filled provisioning gaps that resulted from conflict and/or urban growth.

[111] Counihan, C. M., Around the Tuscan Table: Food, Family and Gender in Twentieth-Century Florence (New York: Routledge, 2004), 119.

[112] Helena Tuomainen, 'Eating Alone or Together?', 7.

[113] Diner, H. R. Hungering for America: Italian, Irish and Jewish foodways in the Age of Migration (Cambridge: Harvard University Press 2001), 73.

[114] Psyche Williams-Forson, 'I Haven't Eaten', 74.

Microclimates on the coast contributed to the conception of Fanteness, Gãness, and Eweness through the foods cooked and eaten there, in relation to each other, to their environments, and their histories. "Food history is a history without names."[115] E. N. Anderson went on to say that it is in the production and consumption of food in interaction with various conditions of soil, water, and weather that people construct their identities. Growing, cooking, and eating the unique foods found in each microclimate allowed people to inhabit a place. That sense of place took up residence and was embodied through the fleshly interaction with food's aroma, feel, sound, taste and smell. These processes endowed the body with a sense of place and provide it with a detailed understanding of that place.[116] This meant that for many coastal peoples, being and personhood were profoundly relational and in conversation with other beings, gods, animals, and environments.[117] The relational experiences of the Fante, Gã, and Ewe contributed to the staying power of the foods associated with each group, irrespective of the conflict, drought, and plague experienced on the coast in the nineteenth century.

References

G.A. Acquah, *The Fantse of Ghana: A History* (Hull University, 1957)

C.D. Adams, Activities of Danish botanists in Guinea 1783–1850. Trans. Hist. Soc. Ghana **3**(1), 30–46 (1957)

E.K. Akyeampong, *Drink, Power, and Cultural Change* (Heinemann, Portsmouth, 1996)

E.K. Akyeampong, *Between the Sea and the Lagoon* (James Currey Publishers, 2001a)

E.K. Akyeampong, *Between the Sea and the Lagoon: An Eco-Social History of the Anlo of Southeastern Ghana: C. 1850 to Recent Times* (Ohio University Press, Athens, 2001b)

[115] E.N. Anderson, *Everyone Eats: Understanding Food and Culture* (New York, New York University Press, 2005).

[116] Wirzba, *Food and Faith*, 40.

[117] Jean Zizioulas and John D. Zizioulas, *Communion and Otherness: Further Studies in Personhood and the Church* (London: T & T Clark, 2006), 39.

K. Ambah, Fish ovens—destruction of by Accra town council. ADM 11/78. PRAAD Accra, February 1909

K. Ambah, G.M.T. Obiri, Fish ovens—destruction of by Accra town council, ADM 11/1/78. PRAAD Accra, February 1909

E.N. Anderson, *Everyone Eats: Understanding Food and Culture* (New York University Press, New York, 2005)

A. Anyidoho, M.E.K. Dakubu, Ghana: indigenous languages, English, and an emerging national identity, in *Language and National Identity in Africa*, ed. by A. Simpson, (Oxford University Press, Oxford, 2008), pp. 141–157

G. Austin, Between abolition and Jihad: the Asante response to the ending of the Atlantic slave trade, 1807–1896, in *From Slave Trade to 'Legitimate' Commerce: The Commercial Transition in Nineteenth-Century West Africa*, African Studies Series, ed. by R. Law, vol. 86, (Cambridge University Press, Cambridge, 1995), pp. 93–118

BetumiBlog: Ghana-Style Kenkey, https://www.betumi.com/2007/03/ghana-style-kenkey-italy-has-polenta.html. Accessed 18 May 2020

W. Bosman, *A New and Accurate Description of the Coast of Guinea* (Cass & Co., London, 1967)

R.F. Burton, *Wanderings in West Africa from Liverpool to Fernando Po* (Tinsley brothers, London, 1863)

J.G. Christaller, *Collection of Three Thousand Six Hundred Tshi Proverbs* (Edwin Mellen Press Ltd, Lewiston, 1990)

G. Clark, *Onions Are My Husband: Survival and Accumulation by West African Market Women* (University of Chicago Press, Chicago, 1994)

M.J.J. Crooks, *Records Relating to the Gold Coast Settlements from 1750 to 1874* (Routledge, 2013). https://doi.org/10.4324/9781315032672

B. Cruickshank, *Eighteen Years on the Gold Coast of Africa* (London, Hurst and Blackett, London, 1853)

P. De Marees, *Description and Historical Account of the Gold Kingdom of Guinea* (British Academy, Oxford, 1987)

J. Dieterle, Basel Mission Report Aburi (PRAAD, Aburi, Aburi, Ghana, 1858). ABM, D-1,9

E.V. Doku, *Cassava in Ghana* (Ghana Universities Press, Accra, 1969)

M. Douglas, *Purity and Danger: An Analysis of Concepts of Pollution and Taboo* (Routledge, London, 1966)

M. Douglas, *Natural Symbols: Explorations in Cosmology* (Penguin, Harmondsworth, 1973)

R.E. Dumett, African merchants of the gold coast, 1860–1905—dynamics of indigenous entrepreneurship. Comp. Stud. Soc. Hist. **25**(4), 661–693 (1983)

J. Duncan, *Travels in Western Africa, in 1845 & 1846* (R. Bentley, London, 1847)
A.B. Ellis, *West African Sketches* (Tinsley, London, 1881)
M.J. Field, *Social Organization of the Gã People* (Crown Agents for the Colonies, Accra, 1940)
S.E. Greene, *Gender, Ethnicity, and Social Change on the Upper Slave Coast* (Heinemann, Portsmouth, NH, 1996)
C.H. Hart Davies, Kormantine fishermen attack made on by Anomabo fishermen for using Ali nets in their waters. ADM 11/1/80. PRAAD Accra, 16 February 1909
F.M. Hodgson, Gold Coast Annual Report for 1895. Annual. Colonial Reports. London: Parliment, February 1897., http://libsysdigi.library.illinois.edu/ilharvest/africana/books2011-05/5530214/5530214_1895/5530214_1895_opt.pdf
H. Huber, *The Krobo: Traditional Social and Religious Life of a West African People* (Anthropos Institute, St. Augustin, 1963)
W. Hutton, *A Voyage to Africa* (Longman, Hurst, Rees, Orme, and Brown, London, 1821)
F.R. Irvine, *The Fishes and Fisheries of the Gold Coast* (Crown Agents for the Colonies, London, 1947)
P.E. Isert, *Letters on West Africa and the Slave Trade*. Translated by Selena Axelrod Winsnes (Legon, Accra: Sub-Saharan Publishers, 2007)
H. Jeppesen, H. (1966). *Danske plantageanlæg på Guldkysten 1788–1850. Geografisk Tidsskrift–Danish Journal of Geography*, 65(1), 48–88. Retrieved from https://tidsskrift.dk/geografisktidsskrift/article/view/128850
R.A. Kea, Plantations and labour in the south-east gold coast from the late eighteenth to the mid nineteenth century, in *From Slave Trade to Legitimate Commerce: The Commercial Transition in Nineteenth-Century West Africa*, ed. by R. Law, (Cambridge University Press, Cambridge, 1995)
E. Kimberley of. No. 224 the Earl of Kimberley to colonel Harley, C.B. 11. CO 879/4/493. National Archives, Kew, July 1873
J.D. La Fleur, *Fusion Foodways of Africa's Gold Coast in the Atlantic Era* (Brill, Leiden, 2012)
R. Laudan, *Cuisine and Empire: Cooking in World History* (University of California Press, Berkeley, 2015)
R. Law (ed.), *From Slave Trade to 'Legitimate' Commerce: The Commercial Transition in Nineteenth-Century West Africa: Papers from a Conference of the Centre of Commonwealth Studies, University of Stirling*, African Studies Series, vol 86 (Cambridge University Press, Cambridge; New York, 1995)

C. Lévi-Strauss, *Totemism* (Merlin Press, London, 1964)

L.A. Lindsay, "To return to the Bosom of their fatherland": Brazilian immigrants in nineteenth-century Lagos. Slavery Abolit. **15**(1), 22–50 (1 April 1994). https://doi.org/10.1080/01440399408575114

P.E. Lovejoy, Background to rebellion: the origins of Muslim slaves in Bahia. Slavery Abolit. **15**(2), 151–180 (1 August 1994). https://doi.org/10.1080/01440399408575130

M. Lynn, The West African palm oil trade in the nineteenth century and the "Crisis of Adaptation", in *From Slave Trade to 'Legitimate' Commerce: The Commercial Transition in Nineteenth-Century West Africa*, African Studies Series, ed. by R. Law, vol. 86, (Cambridge University Press, Cambridge, 1995), pp. 57–77

G. Macdonald, *The Gold Coast, Past and Present* (Longmans, Green, London, 1898)

D.J.E. Maier, Precolonial palm oil production and gender division of labor in nineteenth-century gold coast and Togoland. Afr. Econ. Hist. **37**, 1–32 (2009)

M. McCarthy, *Social Change and the Growth of British Power in the Gold Coast: The Fante States, 1807–1874* (University Press of America, Lanham, 1983)

P. Mensah, A.M. Tomkins, B.S. Drasar, T.J. Harrison, Antimicrobial effect of fermented Ghanaian Maize Dough. J. Appl. Bacteriol. **70**(3), 203–210 (1991)

H. Meredith, *An Account of the Gold Coast of Africa* (Longman, Hurst, Rees, Orme, and Brown, London, 1812a)

H. Meredith, *An Account of the Gold Coast of Africa: With a Brief History of the African Company* (Longman, Hurst, Rees, Orme, and Brown, 1812b)

H.C. Monrad, *Two Views from Christiansborg Castle*. Translated by Selena Axelrod Winsnes (Legon, Accra: African Books Collective, 2009)

M. Obodai, C. Uduro-Yeboah, W. Amoa-Awua, G. Anyebuno, H. Ofori, T. Annan, C. Mestres, D. Pallet, Kenkey production, vending, and consumption practices in Ghana. Food Chain **4**(3), 275–288 (1 October 2014). https://doi.org/10.3362/2046-1887.2014.027

F. Osseo-Asare, B. Baëta, *The Ghana Cookbook* (Hippocrene Books, New York, 2015)

J. Parker, *Making the Town: Ga State and Society in Early Colonial Accra* (James Currey, Oxford, 2000)

J. Rask, *Two Views from Christiansborg Castle*. Translated by Selena Axelrod Winsnes (Legon: African Books Collective, 2009)

C.C. Reindorf, *History of the Gold Coast and Asante* (Missionsbuchhandlung, Basel, 1895)

J. Rich, *A Workman Is Worthy of His Meat: Food and Colonialism in the Gabon Estuary* (University of Nebraska Press, Lincoln, 2007)

J.E. Robins, "Food Comes First": the development of colonial nutritional policy in Ghana, 1900–1950. Glob. Food Hist. **4**(2), 168–188 (2018)

U. Sill, *Encounters in Quest of Christian Womanhood* (Brill, Leiden, 2010)

Supreme court fishing nets decision riots. ADM 11/1/628. PRAAD Accra, 1916

L.E. Wilson, The "Bloodless Conquest" in Southeastern Ghana: the Huza and territorial expansion of the Krobo in the 19th century. Int. J. Afr. Hist. Stud. **23**(2), 269–297 (1990)

N. Wirzba, *Food and Faith: A Theology of Eating* (Cambridge University Press, 2011)

J. Zizioulas, J.D. Zizioulas, *Communion and Otherness: Further Studies in Personhood and the Church* (T & T Clark, London, 2006)

6

Savanna Foodways

On my first visit to the north of Ghana in 2014, I was told by people in Accra that northern Ghana was different from the south. When I asked why the north was different, many attributed this difference to the larger Muslim population in the north. John Mahama, a member of the northern Gonja ethnic group, was president at the time, and there was considerable pride in the inclusiveness that his presidency represented. After a sixteen-hour bus ride to Bolgatanga, and subsequent travel to places like Tamale and Mole, I concluded that indeed, the north of Ghana was different especially with respect to its food. There were some similarities that the food of northern Ghana shares with the south and indeed with the rest of West Africa; the main meal based on a starchy staple accompanied by a soup. However, the savanna climate shaped the way people approached cooking and eating in ways that were distinct from the south of Ghana.

The climatic difference in the north is due to the difference in rain patterns. In his work on the history of foodways in Africa, James McCann emphasized the seasonality of African cuisine:

B. Simpson Miller, *Food and Identity in Nineteenth and Twentieth Century Ghana*, Food and Identity in a Globalising World,
https://doi.org/10.1007/978-3-030-88403-1_6

> Africa's physical environment and climate imposed on it a distinctive annual rhythm—that of the seasonal calendar of climate, movement, and human ritual. In the temperate zones of the world, climate, growing seasons and cycles of life respond most directly to fluctuations in temperature, but Africa's rhythms of life reflect the availability of moisture, especially rainfall. The shifting of the seasonal rain-bearing turbulence on an annual basis sets a general two-part pattern of season, one wet and one dry.[1]

McCann goes on to state that the contrast between wet and dry seasons in Africa is "the most unforgiving of any continent". McCann's emphasis on seasonal moisture is especially useful as it provides insight into the differences between Ghana's northern savanna climate and its southern coastal and forest biomes. The difference in seasonal moisture modified ideas about what the proper meal looked like in northern Ghana. This chapter then closely reviews the climate, available foods, settlement patterns, and techniques of cooks in northern Ghana in order to background the differences between the north and the south.

Historiography

Pre-colonial information about the foodways of the Northern Territories is scarce, but there are a few sources that identify the staple foods in the area. In 1817, Bowdich observed that hundreds of sheep and cows were paid as tribute to the Asante from the savanna kingdom of Dagbon.[2] Fante Gold Coast government official and cartographer George Ekem Ferguson (1890–1897) provided information about the staples of northern cuisine in the late nineteenth century from a map of northern intra-regional trade products that included shea butter, groundnuts, yams, and cassava.[3] During the same time period, French officer Louis Binger (1892) observed staples of maize, yam, and cassava in the market town of

[1] James C. McCann, *Stirring the Pot: A History of African Cuisine* (Athens, Ohio: Ohio University Press, 2009), 17.

[2] Bowdich, *Mission*, 320–21.

[3] Kwame Arhin, ed., *The Papers of George Ekem Ferguson: A Fanti Official of the Government of the Gold Coast, 1890–1897* (Leiden: Afrika-Studiecentrum, 1974), 178.

Kintampo. Millet and sorghum he considered supplementary and costly: "It was not possible for me to buy this last cereal and the other commodities were very expensive."[4] Binger also noted that salt, beef and shea butter were fetched from Salaga further northwest, these being the other foundational items of the northern diet. Yam too was important as a staple but was typically grown further northwest and exported to entrepots like Kintampo: "The Gurunshi have yam plantations. They don't eat the yam but sell it."[5] These early observations help develop an understanding of the variety of starches in nineteenth century northern Ghana, and hint at their seasonal availability.

The region became a focus for ethnographic research, beginning in the 1920s with British officials A. W. Cardinall and R. S. Rattray, and continuing with the first generations of professional social anthropologists: Meyer Fortes in the 1930s and Jack Goody in the 1950s. Food supply and nutrition in the domestic economy were an important part of intelligence gathering in the administration and in the construction of dependencies.[6] The work of these scholars touched on the issues of foodways, nutrition and scarcity in the northern savanna and represent the most comprehensive historical information available about the foodways of the north. Consequently, much of the available information on the foodways in the north of Ghana is derived from early twentieth century anthropological studies and was focused on specific ethnicities. However, it is possible to extrapolate what people were cooking and eating in the region just before the colonial period from a few nineteenth century accounts, early colonial records of market activity, and the archaeological research done in the Banda region of Ghana by Amanda Logan in 2012. These sources revealed that the savanna climate dictated the seasonal variety of the foodways, irrespective of the political configuration or ethnicity of the people inhabiting the region.

---

[4] Louis Gustave Binger, *Du Niger au Golfe de Guinée par le pays de Kong et le Mossi, par le capitaine Binger (1887–1889)* (Hachette, 1892) 155–57.

[5] Binger, *Du Niger,* 28.

[6] Jean Allman and John Parker, *Tongnaab: The History of a West African God* (Bloomington: Indiana University Press, 2005), 57, 73.

Topography

The conclusion of the Anglo-French Convention of June 1898 established the 11th latitude north as the northern boundary of the Northern Territories Protectorate.[7] The British transition to civil administration began officially in 1907. The broad social and political configuration of the savanna regions of the middle Volta basin was characterized by a number of states, the so-called Mossi- Dagomba kingdoms (such as Dagbon) plus Gonja, on the one hand, and non-centralized 'stateless' societies, such as the Tallensi and the Dagara on the border with present day Burkina Faso in the north and Côte d'Ivoire in the west. There are forty ethnonyms that makeup the population in the north that speak a variety of related languages, but still exhibit considerable cultural similarities. The three primary ethnic groups are Dagbani of the Northern Region; Dagaare of Upper West Region; and Frafra—which includes the Nankani, Talni, Bulsa, Nabit, Kusaal, and Grusi (or Grunshi) of the Upper East Region.[8]

Settlement patterns resemble layers, with stateless societies concentrated in the north and centralized societies concentrated further south in the Volta Basin. Just north of Ghana, Burkina Faso was dominated by the Islamic Mossi states of Yatenga and Wagadugu. As one traveled south to Ghana, Mossi villages disappeared into compound type dwellings populated by stateless societies which straddled the border between Burkina and Ghana, separating Mossi states in Burkina from Mamprusi and Dagomba states further south in Ghana (Fig. 6.1).

The population concentrations of these disparate ethnic groups varied greatly throughout the region. The Volta River system drains most of the present-day countries of Burkina Faso, Ghana, parts of Togo, and Côte d'Ivoire. The vegetation changed from north to south, increasing in rainfall and length of wet season the further south one travelled. The Gambaga scarp in Ghana marked an ecological division—south of the scarp yam

[7] A. A. Iliasu, 'The Establishment Of British Administration In Mamprugu, 1898–1937', *Transactions of the Historical Society of Ghana* 16, no. 1 (1975): 2; R. B. Bening, 'Land Policy And Administration In Northern Ghana 1898–1976', *Transactions of the Historical Society of Ghana*, no. 1, vol. 16, 2 (1995): 228, 234.

[8] Awedoba, 'The Peoples of Northern Ghana', 4.

Fig. 6.1 Compound outside Bolgatanga

was grown on a large scale along with millet, maize and sorghum. According to the 1921 census, population densities in the north varied significantly; above the scarp in north-eastern Ghana the population density was 82.2 persons per square mile. In the centre-north the density dropped significantly to only 9.3 inhabitants per square mile. In the far northwest, Lawra's population density was 53.6 per square mile. As savanna approached the forest further south, both dry north crops and wet south crops could grow. Nevertheless, this was a liminal zone where neither type of crop thrived, and the population density dropped. In the 1950s in Gonja nearer the forest zone the population density was only two to five persons per square mile.[9]

The density of settlements and the use of the land in the north of Ghana were dictated by the varying levels of fertility of the microclimates there. There are two major soil groups in northern Ghana: sandstones of the Volta Basin and granites on the plateaus. The centralized states of

[9] Thomas Eric Hilton, *Ghana Population Atlas: The Distribution and Density of Population in the Gold Coast and Togoland Under United Kingdom Trusteeship* (Edinburgh: T. Nelson, 1960).

Dagomba, Mamprusi and Gonja inhabited the Volta Basin.[10] The Basin extends from western Gonja (which encompasses present day Mole National Park) and Wa, to the northern Mamprusi plateaus in the northeast, to the Mampong scarp (some 60 km northwest of Kumasi) in the south. The land was less fertile, flooded during the rainy season and experienced excessively rapid runoff in the dry season. This affected settlement patterns in the area, which were characterized by low population density. Settlements were composed of small villages surrounded by bush farms one to five miles from the settlement, many of which lay fallow for extended periods of time due to low soil fertility. Cattle were kept on grazing land, which was punctuated by trees and farms. In the northeastern part of the basin, cereals were the major crop. In the southeast section, yams are grown on newly cleared bushlands and are then replaced with cereals.

The plateaus above the Gambaga scarp by contrast were characterized by gently rolling hills with heavier rainfall and less permeable soil. As the farming conditions improved above the scarp, these areas were densely populated with compound farms. Compound farms were composed of culturally distinct tribal groupings of indigenous peoples, also known as stateless societies. These farms were under continuous cultivation and fertilized by household and animal waste near the compound/home. Granaries were a prominent architectural feature as physical expressions of the family's economic viability, grain being the foundational staple in these regions as yam did not grow as readily. [11]

Regional Microclimates & Available Foods in the Late Nineteenth Century

The climatic regime of the plateaus and the Volta Basin varied greatly between sub-humid and semi-arid, annual rainfall ranged from 100–150 cm in the Tamale area, falling to 69–119 cm in the Upper East.

[10] Labelle Prussin, *Architecture in Northern Ghana: A Study of Forms and Functions* (Berkeley: University of California Press, 1969), 10–14.

[11] Prussin, *Architecture*, 51–54.

Fig. 6.2 Dawa-dawa seeds in Tamale market 6/10/2014

Rainfall in a six-month season from April to September was patchily distributed, and farmers planted in two phases: early millet followed by late millet and sorghum. This precipitation pattern supported dense growth of tall or short grasses, but not forests (Fig. 6.2).

Regardless of the absence of forests, the seeds and fruits of trees played an important role in the foodways of the north. The dominant tree species were locust bean (*dawa-dawa*) and shea (*Parkia biglobosa*). Further north, the baobab (*Adansonia digitata*) was also found.[12] *Dawa-dawa* tree pods yielded seed balls, and a floury powder made from the pulp provided an umami flavour to soups. *Dawa-dawa* aided in the uptake of protein from staple starches and was itself high in protein and fat. It was

[12] Blench, *Natural Resource*, 21–26.

thought to aid in the creation of blood.[13] Seeds of the *dawa-dawa* pod were boiled, loosened by pounding with a mortar, sieved, boiled again, and fermented, which helped the seeds to develop a highly prized mucilaginous texture that was found to be satisfying. It could then be used fresh or left in the sun to dry. In its raw form processed *dawa-dawa* was reminiscent of a ripe soft cheese and it was served crumbled into soups. Like shea butter, women sold *dawa-dawa* at local markets for extra income as it was labour intensive to produce (Fig. 6.3).

Another important tree product that was resistant to plantation cultivation, shea nuts were crushed to produce a thick yellow cream that has been commercially traded since medieval times.[14] Shea butter, as important to northern cuisine as palm oil was to the south, was traded south to the coast in the very early twentieth century in exchange for items like salt.[15] Shea butter was an essential ingredient in soup, especially during the dry season.[16] The very name of the city of Tamale was etymologically derived from the Dagbani *támá*, meaning shea fruit, an indication of its importance for trade in the region.[17]

The importance of the shea tree meant it was tied to religious rights and that the trees were closely managed. The ritual calendar was organised around the maturation of shea fruit, which occasioned the beginning of the season of plenty and initiated intra and interclan socialisation.[18] By observing the consumption behaviour of fruit bats, who signal the maturation of shea fruits in June, the Dagara (Dagaaba) people knew when to initiate Bagre, a ritualised environmental narrative and performance, and

[13] Jessica R. Ham, 'Cooking to Be Modern but Eating to Be Healthy: The Role of Dawa-Dawa in Contemporary Ghanaian Foodways', *Food, Culture & Society* 20, no. 2 (2017): 244.

[14] Brenda Chalfin, *Shea Butter Republic: State Power, Global Markets, and the Making of an Indigenous Commodity* (Milton Park: Routledge, 2004), 1.

[15] Graham, M.D., W. 'Report on the Vegetable Foodstuffs in Use at Kintampo', Annual Report, Gold Coast, N. T. (1902), ADM 56/1/458, 8.

[16] Andrew Wardell and Niels Fold, 'Globalisations in a Nutshell: Historical Perspectives on the Changing Governance of the Shea Commodity Chain in Northern Ghana', *International Journal of the Commons* 7, no. 2 (August 29, 2013): 367–405.

[17] Roger Blench, *Dagbani Plant Names* (Cambridge: Kay Williamson Educational Foundation, 2012), 19.

[18] J. M. Dalziel and J. Hutchinson, *The Useful Plants of West Tropical Africa* (London: Crown Agents for the Colonies, 1937), 353.

Fig. 6.3 Shea butter at Tamale market

a time for entry into important rites of passage.[19] These valuable trees, found both in the bush and preserved on farms, were traditionally managed by the earth priest or *tendaana* (or by the head of the compound) and could only be harvested or cut with their permission.[20] Cereals in the north provided most of the caloric intake after the wet season, supplemented by legumes and roots. Millet was the most important staple crop, followed by sorghum (guinea corn), a vital secondary crop intercropped with millet, pulses, and with groundnuts. Wild foods and tree produce like shea and *dawa-dawa* were valuable sources of non-cereal foods, especially during the dry season (Fig. 6.4).[21]

[19] Dannabang Kuwabong, 'Bagre: A Dagaaba Celebration of Environmental Balance Between Humans and Non-Humans', *Journal of Dagaare Studies* 4 (2004): 9.

[20] Blench, *Natural Resource*, 31.

[21] Destombes, 'Nutrition', 123.

Fig. 6.4 Women farming millet in Bolgatanga

A long-serving official in the Northern Territories, Cardinall published an analysis of the northern microclimates in his 1931 *The Gold Coast*. In this monograph, he paid close attention to the staple foods of each zone. In this book, he divided the colony into zones A to F from the eleventh parallel border with French Upper Volta (present day Côte d'Ivoire), to the coast. Zones A through C covered the Northern Territories. Zone A encapsulated Navrongo in the northeast, and was dependent on millet, guinea corn (sorghum), beans, groundnuts, and shea butter as staples. Wa and Lawra in the far northwest were reference points for zone B, whose staple foods and proteins were equivalent to those in zone A "but yams are now frequent, and maize occurs".[22] This early twentieth century data corroborates the late nineteenth-century seasonal variety of available staples in each zone.

In Zone C (the area from Salaga down to Wenchi) located just north of ten degrees latitude and inclusive of the southern section of Mamprusi, yams became the main starchy staple, augmented by maize, millet, beans, and sweet potatoes. Proteins were equivalent to those in zone A, but in

[22] Cardinall, *The Gold Coast, 1931*, 230.

"considerably less quantities". All three zones, A through C, retained shea butter and groundnuts as their main source of oil. (Cassava was notably absent from this assessment.) What these comparisons of data make clear is that the more remote from the forested region a settlement is, the more heavily meals were composed of millet and guinea corn augmented by beans, groundnuts, and shea butter. The closer a settlement was to the forested area, the more likely it was that yam was the main starchy staple.

The Coming of Islam

Islam has been a factor in the social organization of northern Ghana for centuries. Muslim traders became closely linked with the political organization of centralized states that inhabited the Volta Basin, beginning in the late fourteenth century. This area contained some of the most important markets for trade to the north between Kumasi and the Sudanic zone and represented the coming together of the food cultures of northern savanna and the southern forest and coastal climates. The Volta Basin was intersected by trade routes for gold and kola from the forest. Islam came to northern Ghana as a result of trade: "Analysis reveals…the existence of a Wangaran substratum and a Hausa overlay. In historical terms, the former is to be linked with the ancient trade in gold to the north-west, and the latter with the more recent trade in kola to the north-east."[23] The first wave of Wangara traders came from the northwest from the eleventh to the sixteenth century. The second wave of Hausa from the northeast came in the nineteenth century.[24]

Islam's relationship with locals was one of dispersion and of integration. Traders tended to settle among centralized states at the invitation of chiefs—this is the first phase of the dispersion of Islam. In the second phase Muslim guests were integrated via the building up of communities alongside hospitable chiefs. Over time, elements of Islamic tradition were incorporated into the state. Muslim holy days were celebrated by the

---

[23] Ivor Wilks, 'The Growth of Islamic Learning in Ghana', *Journal of the Historical Society of Nigeria* 2, no. 4 (1963): 410.

[24] Nehemia Levtzion, *Muslims and Chiefs in West Africa: A Study of Islam in the Middle Volta Basin in the Pre-Colonial Period* (Oxford: Clarendon Press, 1968), xxv.

chief and Muslims were integrated into the socio-political system of state. Their presence became a requirement for the enskinment (installation) of a new chief. Over time, Islamic beliefs would diffuse to locals. Many chiefs welcomed Muslim traders as they each justified the other's existence. Muslims provided rituals and learning which acted to validate the power of the chief. Often chiefs would marry their daughters to Muslim traders who would in turn become the new chiefs.[25]

Although the presence of Islam in the north was pervasive, it was not universally embraced. There are variations in the degree to which Muslims were integrated into Mossi-Dabomga states (Dagomba and Mamprusi) and in Gonja, depending upon whether Muslims were welcomed by individual chiefs. Among the stateless societies, Islam largely failed to diffuse. The Nakanse, Kasena, Isala, Awuna, and Tallensi are thought of as Grunshi peoples in the Ghanaian south (although not all these groups accept this moniker). Islamic elements there had not penetrated the culture as thoroughly as in the organized states with chiefs. Muslims and Grunshi people traded with each other, and Muslims were consulted for medicines, but among the Grunshi, Muslims tended to live as separate communities.[26] The notable exception to this would be among those Grunshi who were implicated in the revival of the slave trade in the last quarter of the nineteenth century.

As important as Islam was to the political maintenance of the centralized states in the north of Ghana, little information about its effects on the local foodways can be found in the historical record. A closer look at their influence in the region proves that Islam had a greater influence on trade and politics than on the foodways. Rainfall patterns proved to be a greater determinant of day-to-day cooking and eating than the practices of the Muslim faith. A poem entitled "Poverty" written in the early colonial period (1923) by al-Hujj 'Umar of Salaga, demonstrated the local Muslim preference for wheat, meat and rice over local foods. It was thought that a diet of wheat, meat and rice indicated wealth:

[25] Ibid., xxii, 21, 69, 111.

[26] Ibid., 143.

> In times of famine the poor man's food is the leaves of jute, the locus [locust gum] tree and *gasaya* (a bush). Cassava, *taura*, *gaude*, and *tsada* (three wild fruits), and *dinya* (a black berry), *tafasa* (a bush), *and rujiya* (a wild tuber), and *faru, gonda* (wild fruits)—and shea-fruit and wild water melon, real marks of poverty… Even during a famine, the rich man's diet remains the delicious *taliya* (a wheaten pasta similar to spaghetti), and *kaki* (a wheat savoury pancake), rice cake with rich gravy, soft thick porridge mixed with rich gravy, *kuskus* (steamed crushed durum wheat semolina) and *chacchacka* (meat on a skewer) mixed together and bathed in a rich juicy stew.[27]

Though this poem communicated the sense that cassava and gathered local foods were considered the foods of the destitute, Muslim influences on local foodways were bound by the demands of the local climate. There was mention of the historically Muslim influenced adoption of tea and cakes in a reflection on change over time in consumption patterns in the north of Ghana.[28] As regards Muslim holidays, Goody found that food and drink prohibitions during high holy days like Ramadan were observed by only the strictest adherents.[29]

Among observant Muslims today what is cooked and eaten is representative of the diets of most northerners. A recent study of Tamale adolescents fasting for Ramadan revealed that their daily meals were characteristic of what was expected in the region. Students took tea with bread for their early breakfast, a characteristic breakfast repast for many people in Ghana today regardless of region. Their 'heavy meal' consisted of rice with stew or *tuo zaafi* with soup—fare typical for most northerners.[30] There is little historical evidence of any significant difference in the day-to-day foodways between Muslims and non-Muslims in the nineteenth and early twentieth century northern Ghana.

---

[27] Umar b. al-Hajj Abi-Bakr, 'Poverty', trans. I.A. Tahir, 1923, IASAR/371, Institute of African Studies, Legon.

[28] Jack Goody and Esther N. Goody, 'Food and Identities: Changing Patterns of Consumption in Ghana', *Cambridge Anthropology* 18, no. 3 (1995): 7.

[29] Jack Goody, *Cooking, Cuisine and Class: A Study in Comparative Sociology* (Cambridge: Cambridge University Press, 1982), 78.

[30] Zakari Ali and Abdul-Razak Abizari, 'Ramadan Fasting Alters Food Patterns, Dietary Diversity and Body Weight among Ghanaian Adolescents', *Nutrition Journal* 17, no. 1 (11 August 2018): 11.

It is in festivals linked to Muslim ceremonies like Damba (the birthday of the profit) where most historical discussion of Muslim influence on local foodways is to be found. Observances of major Islamic festivals became part of the local culture alongside the traditional seasonal calendar. The three-day Damba festival began with a rice picking ceremony—the visible cleaning and consumption of rice. The second day of the festival, a bull was slaughtered in accordance with Islamic ritual. These activities on the first two days signalled the ruling estates affiliation with Muslims. The consumption of rice was tangible recognition of the migration of Muslims from rice producing locations. This sort of performative cooking and eating was a confirmation of Muslim loyalty to the secular ruler. The subsequent communal meal of Damba porridge (*tuo zaafi*) on the third day confirmed reunion between all the political divisions and subgroups of the state.[31] It was the communal consumption of traditional porridge—not rice—which signified loyalty to the chief and represented the interconnectedness and unity of society. It was in fact local traditions and local foodways that took centre stage in festivals, whereas Muslim traditions and foodways served a supportive function in the performances of state.

Seasonality and The Proper Meal in the Late Nineteenth and Early Twentieth Centuries

In the 1910s–1920s, A. W. Cardinall served as a district commissioner among the Builsa, Nankane, and Kassena peoples in the north-east of the Northern Territories. While there, he spent time studying political organisation, daily life, and religion. He relayed the following story about the Supreme Being who lived in the sky: "The Kassena relate that in the beginning, the sky was close to the ground. An old woman was about to cook, but the sky was in the way, so, in her temper, she cut off a piece and

[31] Carola Lentz, 'Local Culture in the National Arena: The Politics of Cultural Festivals in Ghana', *African Studies Review* 44, no. 3 (December 2001): 49; Esther N. Goody, *Contexts of Kinship: An Essay in the Family Sociology of the Gonja of Northern Ghana* (Cambridge: Cambridge University Press, 1973), 124–25; Goody, *Cooking*, 91; Goody and Goody, 'Food and Identities', 5.

made it into soup. The sky, angered, went away to its present place."[32] Thus in this story—which is similar to the Akan "mortar and pestle" myth of the withdrawal of Onyame recounted in Chap. 3—we learn of the centrality of soup to the diet of people in the far north, the role and character of "woman" in preparing the soup, and the Kassena (and general West African) view of the Supreme Being as remote and distant.[33]

***Tuo Zaafi* (TZ or Tzed)**

1 cup millet flour
¼ to ½ tsp of salt
2 cups cold water

- In a saucepan whisk the water, millet, and the salt until smooth
- On medium heat, cook stirring constantly with the whisk until it thickens
- Switch to a wooden spoon and cook another 5 minutes stirring constantly
- Lower heat, cover, and let cook another 10–15 mins, checking to prevent scorching.
- Place into bowl and ladle on soup. Shape into individual balls and wrap in cling film if serving later to prevent crust forming. Can be served the next day, cut into slices.

Adapted from Osseo-Asare and Baëta. *The Ghana Cookbook*, 182–183.

This myth elucidates the obligation of women in the north to cater for soup ingredients as part of the conjugal contract, men being providers of the starchy staples like rice, sorghum and yam.[34] Further, the myth illustrated the central role of woman as cook and preparer of the proper meal. A woman's role as cook was a service performed for others, and an essential part of the social contract. In a northern polygamous family, mothers usually cooked for sons. When a mother was no longer able to cook for her son, the head wife would cook for her husband and mother-in-law. Older women could impose upon their social inferiors to cook for them and/or to aid them in cooking for a husband or brother. Part of being a woman in northern Ghana meant mastering the art of cooking.

---

[32] Cardinall, *Natives*, 23.

[33] James O'Connell, 'The Withdrawal of the High God in West African Religion: An Essay in Interpretation', *Man* 62 (1962): 67–69.

[34] Goody, *Contexts*, 52.

Similar to the proper meal in the south of Ghana, the foundation of a proper northern meal in the nineteenth century was a starchy staple—in this case millet or sorghum porridge, thick enough for a stirring stick to stand in—accompanied by a soup. Another name for millet porridge was *tuo zaafe* (now known simply as TZ), which is thought to have been introduced by the arrival of the Nafana from present day north-eastern Côte d'Ivoire in the early eighteenth century.[35] *Tuo zaafe*, eaten as a staple from the border with Burkina Faso to the present day Brong Ahafo Region, meant "very hot" in Hausa and Dagbani.[36] Often fermented before cooking, it tended to have a soft consistency or a loaf-like composition (like Italian polenta), with a mild flavour and a velvety texture.[37] In areas where yams were available, TZ was eaten during the wet season from roughly April to September, and yam fufu was eaten from September to May, during the dry season. Today porridge can be made from the traditional millet and/or sorghum, as well as maize, cassava, or a mixture of each depending upon what part of the harvest cycle one was in (Fig. 6.5).

A chapter about food in northern Ghana could hardly be written without some mention of the cooking and eating of yams as part of the proper meal in areas of the north where they were cultivated. African yams have bumpy, though outer skins that look somewhat like tree bark. In the transitional savanna zone where yam grew readily—nearer rivers and the forested central area—yam was a preferred food and was also grown for export to Asante and the coast. The fact that yam production was labour intensive did not seem to diminish their prestige or popularity. Freeman confirmed the prevalence of nearby yam plantations to the north of Bonduku in the late nineteenth century.[38] Yam fufu was the primary nourishment from the beginning of the yam harvest sometime in August and September, until around the time the wet season began and yams ran

[35] Logan, 'History', 318.

[36] Jonathan Brindle, *A Dictionary and Grammatical Outline of Chakali* (Berlin: Language Science Press, 2017), 90.

[37] Fran Osseo-Asare and Barbara Baëta, *The Ghana Cookbook* (New York: Hippocrene Books, 2015), 182.

[38] Richard Austin Freeman, *Travels and Life in Ashanti and Jaman* (Westminster: A. Constable & Co, 1898), 351.

Fig. 6.5 Yam truck in Accra

out. *Tuo zaafe*, or *kambɔ* (Nafaanra/Gonja) porridge made from grain and cassava flour was eaten throughout the wet season.[39]

Yams were not only prepared as fufu; they were roasted in a fire or prepared as "little white cakes". These "cakes" were made with moistened yam flour, agitated in a calabash and served with an "excessively strong and spicy" brown sauce, or meat reduction.[40] Yam was an important part of the meal in Kintampo in the early twentieth century: "The Fantu eat potatoes and the Hausa eat bread", meaning the Hausa processed grain into "kanki" (*kenkey*), while the Fantu (not to be confused with the coastal Fante) subsisted on yams made into *fufu*.[41] Further north in Gonja, yams were prepared boiled or pounded into fufu. As in other parts of the north, yams were consumed for part of the year, and grains and cassava with soup and fish or bush meat at other times of the year.[42]

---

[39] Logan, *History*, 78.

[40] Louis Tauxier, *Le noir de Bondoukou* (Paris: E. Leroux, 1921), 133.

[41] Graham, Report on Vegetable Foodstuffs, 7.

[42] Goody, *Cooking*, 49–87.

Wasawasa—steamed yam flour/yam balls

2 cigarette tins yam flour (yam heads are used)

1. Place a perforated pot over a pot containing boiling water. Arrange sticks and sponge in the upper pot.
2. Wet the yam flour. Put a little at a time in a calabash and shake the calabash from side to side until the flour forms small beads.
3. Sprinkle the mixture on the sponge and steam for one hour.
4. Serve with gravy.

Adapted from Alice Dede—*Ghanaian Favourite Dishes*, p.52.

Yams were part of the diet in some areas of the north for part of the year, but the preparation of millet or sorghum porridge was a foundational part of the diet of all populations in northern Ghana.

As stated, the processing and cooking of food in the north was, and remains, the preserve of women. However, as millet cooked more quickly (10 minutes) than the pounding of fufu or the processing of kenkey, and regardless of the microclimate most northerners ate TZ for a portion of the year, one is tempted to ask how the cooking and preparation of the ideal meal in the north differed from the south? If the emphasis was not on the multiphasic processing of starch, then how did women spend their time? It was the woman's contribution of soup, structured by the seasonality of the climate, which regulated a woman's labour. Even today, in the savanna zone women may be more likely to concentrate on non-staple crops, especially vegetables, leaving the staple crop production to her husband.[43] The differing seasonality of the foodways in the north dictated that women spent more time growing, preserving, gathering, processing, and storing soup ingredients than in processing starch for a meal.

Near compound settlements, women spent many hours in the gardens where they grew okra, melon, gourd, and sweet potato. In the Volta basin where there tended to be nucleated villages with bush farms a distance away from the home, fenced off gardens were kept near the house or at the periphery of the village where groundnuts, okras, tomatoes and

[43] Cheryl R. Doss, 'Men's Crops? Women's Crops? The Gender Patterns of Cropping in Ghana', *World Development* 30, no. 11 (1 November 2002): 6, 12.

garden eggs were grown.[44] In the early twentieth century women kept gardens where they grew two varieties of tomatoes (regular and cherry) that were used crushed in sauces. Okra was also grown for use in sauces, as well as long red peppers, and, "Women still grow onions, but they do very little."[45] This observation by Tauxier reveals that by the early twentieth-century the demand for onions was so great that men began to farm onions as a cash crop for distribution to southern Ghanaian markets. Soup ingredients—tomato, okra, onion—along with garden eggs (African eggplants) became foundational to most companagium in Ghana at this time.[46] Throughout the nineteenth century, a shift in demand affected the gendered production of soup ingredients.[47] Traditionally however, produce grown by women for use in soup was processed, dried, packed and stored in preparation for the dry season.

In addition to garden work and the preservation of produce, the growing and harvesting of groundnuts was a traditional occupation for women. A woman controlled both the crops of groundnuts and other produce and received the proceeds from the sale of their surplus in the market.[48] Groundnuts provided oil for cooking and were a reliable crop that tolerated poor soils and drought and were pest resistant as the pods grew in the soil.[49] Sold in the Kintampo market in 1902, groundnuts were prepared in a variety of ways: "When roasted and beaten to flour, it becomes the favoured food of the Fantes and Accra…it makes an excellent soup appreciated by European and negro alike."[50] Though associated with West African soups and stews, groundnuts were most often consumed boiled or as a snack. Groundnuts served as an essential soup ingredient during the dry season and were an essential backstop in the strategy women adopted to cope with the seasonality of the savanna climate.

[44] Prussin, *Architecture*, 12, 23.

[45] Tauxier, *Le noir de Bondoukou*, 135.

[46] Logan, 'History', 46; Osseo-Asare, 'We Eat First', 49–57.

[47] Padmanabhan, 'The Making and Unmaking', 6; Doss, 'Men's Crops?, 2.

[48] M. Fortes and S. L. Fortes, 'Food in the Domestic Economy of the Tallensi', *Africa* 9, no. 2 (1936): 245.

[49] E. V. Doku, 'Root Crops in Ghana', *Ghana Journal Science* 6 (1966): 41.

[50] Graham, 'Report', 7.

Not only were northern women absorbed with food production in their gardens and the growing of groundnuts, they spent much time in the collection and preservation of gathered foods such as shea nuts, elm leaves, and *dawa-dawa* seeds.[51] The processing of essential ingredients such as shea nuts to make the butter and the processing of the *dawa-dawa* seeds into a cheesy, crumbly protein rich soup ingredient was time consuming.[52] The maintenance of a larder of dried soup ingredients such as vegetables, leaves, and legumes for the dry season was critically important to the economy of the family.

Tallensi Ocro Soup—*Zisaaleg/Ma'ana Ziet*

Handful of ground, dried fish or meat
Boiling water
Handful of red peppers
Few lumps of salt
A piece of dawa-dawa-seed ball
Equal parts fresh and dried ocros

Add the fish or meat to a pot of boiling water. Grind the red peppers, salt, and dawa-dawa-seed with a little water to make a paste and add to the water. Crushed the ocros into a pulp on the grindstone and add to the boiling soup. Simmer 30 minutes while stirring from time to time with the stirring stick.

Adapted from Fortes, M., and S. L. Fortes. 'Food in the Domestic Economy of the Tallensi', 274.

This extreme seasonality structured the demands on a woman's time—much time was spent on the preservation of soup ingredients—and it affected the composition of the soup, the components of which varied greatly depending upon the season. Regardless of the season—wet or dry—mucilaginous soups were considered proper food as they provided a feeling of satiety even when the family needed to economize on the starchy staple.

The natives classify soups into two types: *zisaalag* and *zizaayag*- ' slippery' soup and 'coarse' soup. The first is smooth and viscous: 'It slides

[51] Ham, 'Cooking to Be Modern', 244.
[52] Goody, *Cooking*, 69.

down as soon as you put it into your mouth, before the porridge.' Ocro soup is the most representative of this kind. The coarse variety is made of crushed groundnuts, marrow leaves, hibiscus or similar vegetables. 'It sticks in your mouth after the porridge and you have to swallow it down'.[53]

Dried and stored in kitchens after its harvest, okra was important to the production of a satisfyingly slimy quality and therefore was a favourite ingredient for soups in both the dry and wet season. When okra was unavailable, groundnuts provided a fortifying, consolidated texture to soup during the lean season before the harvest. The preparation of a satisfying soup could also include ingredients such as hibiscus leaves or fermented melon seeds. Thus, northern soup ingredients being very seasonal, varied throughout the year. The southern cook by contrast had a more seasonally consistent list of starches, proteins, and soup ingredients at her disposal.

Protein sources in the Northern Territories, like the starches and other soup ingredients, were varied and localised depending upon the microclimate and the presence or absence of the tsetse fly. Livestock were customarily held for trade in the market in exchange for grain in the lean times, reserved and distributed at feasts, or sold on to the south. Garden hunting, a term used to describe hunting for small animals such as lizards or rodents, was prevalent, and a regular source of protein was provided by groundnuts, cow peas (black eyed peas), geocarpa beans (also known as the Hausa groundnut), and fermented *dawa-dawa* seeds.[54] Fish as a protein was highly esteemed, but usually only hunted in groups in preparation for a feast:

Fish (*ziy*) is also prized as a delicate food, and there is always a small supply of dry fish available in the market throughout the dry, and part of the rainy season. The Tallensi fish the pools of the partially dry Volta River communally in the dry season, but with poor success, as far as the average householder is concerned. As a rule fish is bought in the market, like meat, only for special purposes.[55] When available, fish was dried, ground, and added to soups. In most regions of the far north, the only

[53] Fortes and Fortes, 'Food', 266.

[54] Ibid., 276.

[55] Ibid., 249.

meat directly killed for eating was the guinea fowl. All other meats, such as beef, goat (the most common), or donkey meat, were ritually killed and communally dispersed.[56] Men hunted seasonal game animals such as duiker (small antelope—*Cephalophus sp.*), antelope (*Kobus sp.*), or bush-buck. Near rivers, catfish, turtles, or perch would round out the offerings.[57] Accordingly, proteins like starches, were seasonal and more varied than what was on offer in the south.

Another way that seasonality affected the foodways of the north is in the function of markets in the distribution of grain, meat, and prepared foods. As previously stated, the main purpose of livestock—beef, mutton, goats, game, and poultry—was to store household wealth that would then be exchanged for grain during the lean or dry season. [58] Temporary markets facilitated intra-district food flows from areas of surplus to areas of deficit.[59] These markets were the backstop for grain sustenance during the lean season, a cultural norm that was part of coping with the seasonality of food offerings.[60] In addition to preserving cultivated and gathered foods, much of women's time during lean times was occupied with preparing food for sale in markets such as *maasa*-fried millet-cake, or fried bean cakes called *guor* and *kameha*.[61] Women prepared beer from ground, soured grain and sold it either from their homes or in the market for extra income.[62] In contrast with the south where food was traded with an emphasis on the accumulation of profit, in the north foods were brought to market as a stopgap during lean times (Fig. 6.6).

Kitchens too reflected the seasonality of the north, and the need to dedicate time to the preservation of garden produce and meats as dictated

---

[56] Ibid., 35.

[57] Achilles Gautier and Wim Van Neer, 'The Continuous Exploitation of Wild Animal Resources in The Archaeozoological Record Of Ghana', *Journal of African Archaeology* 3, no. 2 (25 October 2005): 198.

[58] Fortes and Fortes, 'Food'; Destombes, 'Nutrition'; Franz Kröger, Barbara Meier, and Steve Tonah, "Conflicts and Consensus Between Migrant Fulani Herdsmen and Mamprusi Farmers in Northern Ghana," in *Ghana's North: Research on Culture, Religion, and Politics of Societies in Transition*, eds. Franz Kroger & Barbara Meier (Frankfurt am Main: Peter Lang GmbH, 2003), 89.

[59] Destombes, 'Nutrition', 184.

[60] Ibid., 58.

[61] Fortes and Fortes, 'Food', 260; Goody, *Cooking*, 70.

[62] Fortes and Fortes, 'Food', 25; Goody, *Cooking*, 72, 77.

Fig. 6.6 Tamale courtyard cooking hearth

by the dryer climate. Tools of the Nankanse kitchen in the present-day Upper East Region were described by Rattray in 1932. There was a rack for drying beans near the entrance of the family compound. Opposite the entrance for the dry-weather grinding-room door, was a raised platform for grinding grain, at which one could stand. Firewood was neatly stacked against the wall, as were storage pots on shelves, which contained provisions such as groundnuts. Grain waiting to be processed was hung by string net from the ceiling, along with calabashes. Storage space was essential to accommodate a variety of dried provisions for reconstituting in soup. In the wet-weather kitchen were three hearthstones arranged below a small smoke hole in the ceiling.[63] Women smoked the fish and meat above the fire, reconstituting it by boiling it in soup.

Like those of the Nankanse, the kitchen tools of the neighbouring Tallensi in the early twentieth century were prioritized for the preservation and storage of food. They included a grinding-room with a low, semi-circular platform (*neer*) with two or three depressions and accompanying grinding stones for crushing whole grains and rough maize into fine flour. The smallest stones were used for crushing seasonings for soup, such as Dawa-dawa seed balls and dried meat. Kitchens were in women's

[63] Rattray, *Tribes of the Ashanti Hinterland*, 248–54.

sleeping rooms during the wet season, and in the open-air courtyard in the dry season. The wet season kitchen fireplace consisted of two stones against the compound wall, a low surface for the cook to sit on facing the fire, a stirring stick, and one or two old and partly broken calabashes. Firewood and water stored in large earthenware pots were kept close at hand.[64] Most compounds had multiple hearths, sometimes associated with distinct domestic units although they were often clustered so that women could socialise while cooking.[65] What these northern kitchen set-ups did share with their southern counterparts was the facilitation of cooking with other women, weather permitting.

In addition to the enjoyment of cooking in community, both northern and southern cooks required the labour of others to produce the proper meal. In both the north and in the south, girls were apprenticed from a young age to learn the art of cooking. Again, there were some nuances to the quality of cooking in the north that were dictated by the seasonality of the climate. A prominent example of this subtle difference is found in the meaning of the word "to marry" in some languages in the north:

> For the Konkomba, the concepts of marriage and cooking are inseparable, and in fact the verbs 'to marry' and 'to cook' are the same word (*mcn*) in Likpakpaln, the Konkomba language. When asked to describe the role of a wife, both men and women defined the role of a wife as that of cook; bearing children was the next most commonly cited expectation of a wife.[66]

As discussed in Chap. 2, the verb to eat in Twi was *di*—which also meant to have sex. In southern Ghana it was hoped that sex would result in the production of children—a woman's security in old age.[67] In the north however, the stress was on the cooking rather than on the production of children.

---

[64] Fortes and Fortes, 'Food', 263.

[65] Ann Brower Stahl, *Making History in Banda: Anthropological Visions of Africa's Past* (Cambridge University Press, 2001), 83.

[66] Kelsey B. Hanrahan, '"Mɔn" (to Marry/to Cook): Negotiating Becoming a Wife and Woman in the Kitchens of a Northern Ghanaian Konkomba Community', *Gender, Place & Culture* 22, no. 9 (21 October 2015): 1329.

[67] Jean Marie Allman and Victoria B. Tashjian, *'I Will Not Eat Stone': A Women's History of Colonial Asante* (Portsmouth: James Currey, 2000), 49.

The northern emphasis on a woman's role as cook influenced the spatial organization of homes as well as families. Unlike their southern sisters, marriages in the north had a greater tendency to be virilocal- meaning that women were required to move to their husband's home when they married and were contracted to cook for the husband's lineage: "a woman is 'our wife'".[68] Her labour, the crops she grew, and the children she produced belonged to her husband's lineage. This contrasts with the south where there was a higher tendency for a woman's labour to belong to a matrilineage. In the north, a husband's lineage was obligated to support a woman into her old age, provided she successfully earned a proper position as a cook in the family. A woman's retirement and security were directly dependent upon how successfully she fulfilled her mandate as cook. An individual hearth must be earned after a second lengthy apprenticeship to the husband's mother or his chief wife. Once she earned a place as a cook a woman had the right to command her social inferiors' labour in the processing of food (fetching of fuel or water), to demand supplies like meat and starches from her husband, and to request vegetables and/or seasonings like *kpalago* (processed *dawa-dawa* seeds) from her co-wives or her mother-in-law. (Customarily, it was senior women who had command of the labour of others to fetch fuel for the preparation of *kpalago*.) The emphasis on cooking and the virilocality of wives in the north was a factor of the drier climate and its historically limited farming economy. The southern economy was historically mixed, and a more flexible economy with fishing, trading, and farming offering more security with regards to food and provisioning. By contrast, the seasonality of the foodways in the north as reflected in the dryer climate and greater focus on farming, meant that there was a critical need to ensure that labour was constantly being dedicated to the growth and preservation of both cultivated and gathered foods.

An enduring relational idiom for people living in both the north and the south of Ghana was (and continues to be) the sharing of cooked food. Food was considered a powerful substance that must be carefully managed. In nineteenth-century northern Ghana, the sharing of cooked food was a metaphor for expressing an intimate, trusting relationship.

[68] Hanrahan, 'Mɔn', 1328.

Therefore, the acceptance of cooked food from a stranger was a dangerous proposition: "This custom of sacrifices seems to have been instituted to protect the leaders and influential people from violent deaths, poison playing a considerable role in this country."[69] The use of food as a vehicle for poison was believed to be part of everyday life. Not so much as a kola nut would be accepted from a stranger: "Sorcerer's medicine can be transmitted to the victim in three ways: in beer, in a kola-nut, or by being placed on a path…The Kokomba do not eat kola-nuts given to them by strangers. They accept the nut, thank the giver, and, later, throw it away."[70] To eat food from a stranger was to expose oneself to all manner of potential calamity. This was why the consumption of cooked food was such a powerful expression of communal confidence.

The preparation of cooked food was a service which women performed outside of their role as wife. A woman could be called upon to cook for others in her role as sister, as a social inferior, or in a community leadership role. Irrespective of their central societal role as cooks, women occupied a social position on both sides of the threshold of security. While women were trusted to prepare and distribute cooked food, at the same time women were also considered to be potentially dangerous. This was because the most common way for a witch to attack those toward whom they harboured ill will was to introduce poison to cooked food. To eat a communally cooked meal therefore required trust as it represented a risk.

The communally cooked meal characterised what anthropologist Claude Fischler called the 'omnivore's paradox', the tension between the danger of the unknown and the need for novelty and variety. This paradox represents the fundamental anxiety in man's relationship to his foods. A cuisine's function—the body of practices, representations, rules and norms based on classifications—is precisely to resolve the omnivore's paradox.[71] To mitigate the danger of communal eating during religious and/or political festivals like the Damba, *ewuritches*—holders of women's chieftainships, considered to be trusted sisters—prepared the Damba rice

[69] Binger, *Du Niger*, 187.

[70] David Tait, *The Konkomba of Northern Ghana* (Milton Park: Routledge, 2018), 233.

[71] Claude Fischler, 'Food, Self and Identity', *Social Science Information* 27, no. 2 (1 June 1988): 279, https://doi.org/10.1177/053901888027002005.

eaten by the community.[72] This act of food-sharing was at once a powerful signifier of a close, dependable relationship and a means to reinforce the political leadership's position. Another example of women cooking in an act of communal service was at the enskinment of a chief. It was the chief's sister who cooked all food eaten by him during the installation period, as "it is your sister whom you can really trust"—*fo suputche e kɔ yerda.*[73] During festivals or life events where cooked food needed to be produced and shared, women were ceremonially elevated to the position of "sister" and trusted cooks. In this manner women were entrusted with the responsibility of facilitating the reintroduction of individuals to the community through the medium of cooked food.

At Table

As in most other regions of Ghana, meals were cooked twice daily: once at noon and again at sunset, with a little extra prepared at the evening meal for the next morning's breakfast.[74] Meals were generally eaten semi-communally in the central courtyard of the home, or some would choose to eat their portion in their own rooms. If eaten in the courtyard, bowls of soup and porridge made of guinea corn (sorghum) or millet were laid out in front of the senior man. Women sat with outstretched legs or upon a four-legged stool in the central courtyard.[75] The influence of Islam is seen in Gonja, where water would be first brought to wash one's hands.[76] Similar to the custom in southern Ghana, whom one ate with was structured by ones status: "Women, girls and young children eat together, in the same units as those on which the preparation of food is based. That is, if two co-wives take turns at cooking, they will eat from one pan, whoever has prepared the meal."[77] As in southern Ghana, senior men tended to eat alone, and other, less senior men customarily ate their food together.

[72] Goody, *Contexts*, 3.

[73] Ibid., 122.

[74] Fortes and Fortes, 'Food' 264, 266.

[75] Goody, *Cooking*, 86.

[76] Ibid., 77.

[77] Goody, *Contexts*, 53–54.

Much as in the south, meat was distributed along a hierarchy. In the common soup, women customarily waited until men had taken their portion before they helped themselves, or women distributed the soup in bowls, giving the larger portions of meat to men. However, children were fed irrespective of their sex when they ate with mothers.[78] In a notable contrast to how the Akan in the central forested region fed their children (a gerontocracy where older males ate the best food and the larger portion of meat), the Tallensi allowed their children to eat their fill before adults even began to eat. Even though the father in the home of the Tallensi ate first, he ate with his children (especially his youngest child), not apart.[79] It was the father's job to oversee the feeding of those children that for one reason or another were not getting enough nourishment.

The distribution of cooked food in northern Ghana, and who was deserving of a share, mirrored that of the south. Cooked food was shared daily by a "family"—several sets of people who ate the food a wife prepared.[80] The people attached to the same pot could include dependents both near and far. For instance, regular obligations to feed the mother or father of the husband or wife or a sibling were closely adhered to. Millet and soup for these dependents was apportioned in a calabash and a young child would be sent to deliver the meal.[81] In this way northern organization of meals was similar to that of Ga and the Fante on the coast: "Their common solidarity resides firmly on the cooking pot".[82] The same organization was echoed in the Gonja metaphor "having one bowl".[83] It was the cooking pot/bowl that was the integrative core of a group of people.

[78] Goody, *Cooking*, 68.

[79] Fortes and Fortes, 'Food', 271.

[80] Goody, *Contexts*, 54, 126.

[81] Goody, *Contexts*, 53.

[82] Margaret J Field, *Social Organization of the Gã People* (Accra: Crown Agents for the Colonies, 1940), 60; Emile V. W. Vercruijsse, Lydi M. Vercruijsse-Dopheide, and Kwasi J. A. Boakye, *Composition of Households in Some Fante Communities: A Study of the Framework of Social Integration* (Cape Coast: Institute of Social Studies, 1972), 11.

[83] Goody, *Cooking*, 89.

Transition to the Colonial Period

The blending of cassava and maize flour into the preparation of TZ in the nineteenth century Banda region was reflective of the changes in the trading environment of the Gold Coast throughout the nineteenth century. The abolishment of the slave trade by the British did not abolish the internal slave trade, as labour was still needed to supply growing demand for legitimate trade items such as palm oil and the expansion of gold mining. By the 1820s, people in Banda fled their homes and became subject to a seventy-year period of dislocation and migration. Both oral history and archaeological research confirmed the increased use of gathered foods and the changeover away from sorghum to maize and cassava.[84] In the 1830s palm oil became firmly established as a commodity of major importance, further contributing to the destabilization of northern populations.[85] Internal slave raids and migrations in northern Ghana forced people to adopt other staples in the nineteenth century, which they folded into their preparation of TZ.

The Asante defeat at the hands of the British in 1874 saw another concurrent rise in internal slave trading for agricultural and domestic work, as well as the production of legitimate trade items.[86] The Zaberma turned to slave raiding among the Grunshi in the late nineteenth century, exchanging slaves for guns in the south and for horses in the north.[87] This revival of the slave trade contributed to the low population density in the central part of northern Ghana inhabited by the Grunshi. As the Zaberma were a Muslim minority among the Muslim Wangara and the Hausa, they turned to recruiting local people as volunteers and as captives. Consequently, these local troops took part in raids on their own villages. Many of those Grunshi raiding with the Zaberma (as well as those sold to Wa and Asante), encountered Islam and converted. During the colonial period these Muslim Grunshi served in the British auxiliary forces and participated in

[84] Logan, 'History', 312.

[85] Brodie Cruickshank, *Eighteen Years on the Gold Coast of Africa* (London: London, Hurst and Blackett, 1853), II, 41–41.

[86] Tauxier, *Le noir de Bondoukou*, 62, 260; Logan, 'History', 312.

[87] Tauxier, *Le noir de Bondoukou*, 116; Levtzion, *Muslims and Chiefs*, 159–60.

seasonal migrations to the south for work in colonial export industries.[88] As stated earlier in the chapter, these Muslim soldiers were eating "Hausa kenkey" prepared from maize in the new trading town of Kintampo in the very early twentieth century. These soldiers, called "Hausa" by the British, no doubt encountered maize on the coast of Ghana, which they adopted and consumed in places where they were stationed.

Throughout the nineteenth century in northern Ghana, people turned to new world crops such as maize and cassava for survival as it was often the only food left to eat after conflict and drought. In part a consequence of the occupation of Bonduku near the present-day border with Côte d'Ivoire by Samori, a British expedition was dispatched to the area in 1896. They encountered considerable hardship and challenges in provisioning their troops:

> In consequence of the occupation of Bontuku by Samory—a slave-raider who devastated the French Sudan for many years—… and expedition was dispatched in February 1896 to the neighbourhood of Bontuku. It found whole districts and towns in Gaman [Gyaman] laid waste and depopulated and provisions so scarce that the only foodstuffs locally available for native soldiers and carriers was, with very few exceptions, dried cassada [cassava], ground and made into flour."[89]

This account demonstrated the importance of cassava flour to the survival of locals in an area devastated by years of slave raiding, just seventeen kilometres away from present day Ghana. The inability of the population to farm opened the door to crops like maize and cassava, cassava keeping underground for long periods of time, and maize having a rapid maturation rate. Cassava was especially versatile as its leaves, plucked from abandoned farms, could be used for soup.[90]

The trend toward the adoption of new world crops like maize and cassava grew as the British gained control over Asante and the northern hinterland in the late nineteenth and early twentieth centuries. Flows of trade and labour were redirected to gold extraction and cocoa export

[88] Ibid., 157.

[89] Cardinall, *The Gold Coast,* 34.

[90] Logan, 'History', 282.

from the forested areas beginning in 1891. Within twenty years Ghana became the world's largest producer of cocoa with a harvest of nearly 40,000 tons. In 1923 the harvest was over 200,000 tons.[91] Over time this contributed to a change in the staple foods of the region. Maize had almost completely replaced sorghum in areas closer to the forest, and cassava was used all over the north. New World crops produced higher yields and required lower labour inputs; a boon as men and women began migrating to work in cocoa production areas.

With the expansion of British administration in the early twentieth century, food in the Northern Territories continued to go through a process of mixture and reformulation. Much as in the south, cassava was considered an inferior food in the savanna region. The 1902 report on vegetable foodstuffs confirms this attitude: "It [cassava] is composed almost wholly of starch and is considered poor coarse food at Kintampo".[92] Pressure on food supplies increased with the required provisioning of British troops and carriers based in Kintampo. As early as 1901, the

Cassava Fufu

Cassava tubers
Grater
Leaves
Calabash
Boiling Water
Mortar and Pestle

- Clean and scrape tuber, then grate on tin grater
- Place the grated mass in a calabash and cover with leaves and allow to stand for 2 or 3 days
- Remove from calabash and beat in a mortar for ¼ of an hour
- Place the mass in a pot of boiling water
- Cut into pieces and eat hot or cold

Adapted from Dr W. Graham, Kintampo Annual Report (1902)

---

[91] Gareth Austin, 'Mode of Production or Mode of Cultivation: Explaining the Failure of European Cocoa Planters in Competition with African Farmers in Colonial Ghana', in *Cocoa Pioneer Fronts since 1800: The Role of Smallholders, Planters and Merchants*, ed. William Gervase Clarence-Smith (London: Palgrave Macmillan UK, 1996), 154.

[92] Graham, Report, 6.

soldiers at Bui required 400 yams per diem or a half a yam per man, causing people to look for low input alternatives.[93]

Cassava played an integral role in relieving the pressure people were experiencing from conflict and the transition to colonial rule. From just across the border in French Gyaman, Tauxier wrote that "Another subsidiary crop less important than maize but the ultimate resource in years of hunger is cassava. It is planted at the same time as yam and is harvested fifteen months later."[94] People began to intercrop yams and cassava to ensure they would have enough to eat. Tauxier went on to described war and a drought in the region from 1918 to 1919. During these years the price of yam doubled and the most important subsidiary crop in years of shortage was cassava. By the early 1920s, cassava had become a staple in the north.

The pressures caused by war, displacement, and drought were further exacerbated by the long-term change in the length of wet and dry seasons. Like cassava, maize began to overtake the production of millet and sorghum in the Volta Basin due to its adaptability to different microclimates. There were four varieties of maize noted in the early twentieth century: two varieties of red, one of white, and one of black.[95] By 1902, the District Commissioner in Kintampo stated that maize was the most prevalent staple food of the "Hausa" population that was grown "on every suitable piece of land in or about the town".[96]

No matter the starch people came to rely upon, local cooks took care to replicate historical grain processing methods to achieve the texture they were accustomed to. The methods used for the processing of maize into what was called "Hausa kenkey" were reflective of techniques specific to northern Sahelian climates used to make millet or sorghum porridge. The hardness of those grains proved insect resistant when stored but made processing a more difficult and multi-step process. Most processers used a wooden mortar and pestle, as one would use to process fufu. Some more sedentary people used heavy grinding stones to further process the hard grain.[97] Both a wooden mortar and grinding stones were

[93] Stahl, *Making History*, 196.

[94] Tauxier, *Le noir de Bondoukou*, 133.

[95] Tauxier, *Le noir de Bondoukou*, 133.

[96] Graham, Report, 3.

[97] J. D. La Fleur, *Fusion Foodways of Africa's Gold Coast in the Atlantic Era* (Leiden: Brill, 2012), 43.

used for processing maize, giving this recipe a texture unique to this area. Although the starches people relied upon in the north changed in the twentieth century from millet and sorghum to cassava and maize due to a combination of conflict and climate change, the processing methods were reflective of the regional identity of the people preparing them.

Post-Colonial Transitions

In the 1980s, a savanna famine made some other changes with respect to staple starches in different parts of the north. Cassava and potato replaced the secondary crop of frafra potatoes in the Upper Eastern Region.[98] In Gonja, maize was not fully adopted until a drought in 1985, near the new capital of Damongo, which forced them to adopt a more "southern diet", adapting to *kenkey* and using more cassava in a dish called *kokonte* (cassava).[99] In the 1990s, ecological change in Brong Ahafo resulted in a switch to the exclusive cultivation of maize, with sorghum as a secondary crop. Maize became a convenience food for migrant workers, who purchased maize flour with cash in places like Banda.[100]

The expansion of maize as a staple continued into the post-colonial period. From the 1950s to the late 1980s, the capital of Gonja, Damongo, became a centre of maize production. An analysis of agricultural practice from 1957 also outlined changes in the farming system of western Dagbon. A survey conducted in 1994–1995 yielded some perspective on this change. The most significant finding was that maize cultivation may have been at similar levels as the pre-Independence era, but the cultivation of sorghum and millet declined over time.[101] Where maize used to be a supplementary crop in the early twentieth century, it became a major starch in the mid-to-late twentieth century.[102] By the 1990s, due to

[98] Blench, *Natural Resource*, 36.

[99] Goody and Goody, *Food and Identities*, 5.

[100] Logan, History, 37–43.

[101] Michael Warner, Ramatu Al-Hassan, and Jonathan Kydd, 'A Review of Changes to Farming Systems of Northern Ghana (1957–94)', in *Natural Resource Management in Ghana and Its Socio-Economic Context*, ed. Roger M Blench (London: Overseas Development Institute, 1999), 87.

[102] Goody and Goody, 'Food and Identities', 5.

ecological changes that resulted in a drier climate, cassava and maize replaced the older crops almost entirely in the Volta Basin.[103]

By the 1990s a combination of market forces and climate changes affected the way people cooked and ate soup in northern Ghana. The practice of outdoor preparation and cooking is still relatively common. What has changed is the production of foods that were soup ingredients now being grown for sale to southern Ghana. Over the last fifty years a shift toward the cash crop farming of vegetables such as tomato, onions, and garden eggs for the markets in southern Ghana has been made. Even foods that served as additional sources of protein in the past began to be farmed as commercial crops. For example, beginning in the early 2000s the peanut was both a subsistence crop and a cash crop, farmed by 90% of families in the north of Ghana.[104] Sources of protein have become a particular problem as people no longer source meat from the wild as they did in years past, and the smoked dried fish from the south is expensive. Cow peas have risen in value due to changing agricultural practices and thus have begun to be reserved for sale in the market, adding to the dearth of protein (Fig. 6.7).[105]

Dawa-Dawa was another major source of protein and an essential soup ingredient which has been adversely affected by the gradual change in the length of the rainy season. Climate change is affecting the "lean season" gap between harvests which is beginning several months earlier than it has typically been understood to begin, shifting from May/June to March.[106] Seasonal changes combined with population pressure has precipitated a shift away from the use of *dawa-dawa* in soups in the Upper East. Soybeans were being used to partially or completely replace *kpalago*, a soup ingredient traditionally made of *dawa-dawa* seeds. Used to provide

[103] Kofi Marfo and Steve Wiggins, 'Changing Farming Practices in the Forest Guinea Savannah Zone of Ghana', in *Natural Resource Management in Ghana and Its Socio-Economic Context*, ed. Roger M Blench (London: Overseas Development Institute, 1999), 67–68.

[104] F.K. Tsigbey, R. L. Brandenburg, and V. A. Clottey, 'Peanut Production Methods in Northern Ghana and Some Disease Perspectives', *World Geography of the Peanut Knowledge Base Website* 9 (2003): 33–38.

[105] Martina A. Padmanabhan, 'The Making and Unmaking of Gendered Crops in Northern Ghana', *Singapore Journal of Tropical Geography* 28, no. 1 (2007): 63.

[106] Jessica R. Ham, '"Who Knows Tomorrow?": Food Insecurity, Distress, and Managing the Future in Upper West Ghana.' (PhD diss., University of Georgia, Athens, 2016), 18.

Fig. 6.7 Outdoor Kitchen in Bolgatanga

an umami flavour, protein, and a mucilaginous texture, this critical ingredient has become a cost factor in a woman's budget due to its recent decline in availability. The problem was severe in the densely populated Upper East Region where a reduction in the number of trees has affected the traditional collection and distribution system.

For Kusasi women near the border with Burkina, soybeans were a neutral food and they did not create issues of age or status. For the Dagomba living in the Volta Basin, the collection and distribution of *dawa-dawa* was highly esteemed by men who administered this procedure and whose political standing in the community was connected with this process.[107] This meant that *dawa-dawa* was an essential un-substitutable soup ingredient. In order to cope with the declining availability of *dawa-dawa*, Dagomba women prepared *kpalago* by supplementing the tree seeds with soybeans and cooking them in such a way as to separate them from *dawa-dawa* seeds while cooking: "We separate it, but we mix it in the same pot. We put the *selentujia* [literally: white men's beans] into a rubber

[107] Padmanabhan, 'The Making and Unmaking', 66.

container, tie it tight, and add it to the dawadawa."[108] This allowed Dagomba women to feel as if they were still preparing *kpalago* traditionally, thus satisfying their obligation to provide proper ingredients to make soup.

Some women in the Upper West were eschewing *kpalago* entirely in favour of using seasoning cubes in their soups. In the last fifty years, people in northern Ghana have come to rely more upon the market for the purchase of things they used to hunt or gather. This presented a problem for women who were obligated to spend their money on the all-important soup ingredients. Younger women in the Upper West were more often choosing seasoning cubes as a fast, cheap flavour substitute for *dawa-dawa* and other sources of protein.[109] Traditionally, younger women petitioned senior women for access to *kpalago* as senior women were customarily the ones that processed this essential ingredient. The ability to replace *kpalago* with purchased seasoning cubes has undoubtedly altered relationships around the production and processing of *kpalago*.

Conclusion

Chapter 1 briefly examined the topography and climate that contributed to the elements of the proper meal. Chapter 2 explored the concept of the proper meal in Ghana. This chapter further examines how these ideas applied to the savanna climate of northern Ghana and highlighted how the seasonality of northern Ghana modified its concept of the proper meal in the late nineteenth and twentieth centuries. To be sure, there were several similarities between ideas of the proper meal in the north and the south. In both places a hierarchy to both cooking and eating was observed. The changeover in staple starches to maize and cassava would be familiar to both north and south, although in the south, there is evidence that this process began much earlier than in northern Ghana (in the seventeenth century). The sharing of cooked food being organized around a single pot was also similar in both climates. However, it is the

[108] Ibid., 67.

[109] Ham, 'Cooking to Be Modern', 251.

variance in rainfall and the resulting seasonality that truly distinguished foodways in northern Ghana from those in the south.

These differences in northern Ghana foodways manifested themselves in the way that northern women spent their time processing and storing of a large variety of cultivated and gathered foods in preparation for a lean season when starchy grains were not in production. The centrality of soup, and a women's dedication to and conjugal obligation in securing soup ingredients, differed from the south where the emphasis was on both the cooking of the soup as well as the laborious processing of the starch for the proper meal. In the drier climate of northern Ghana, the need to produce, gather, and store soup ingredients meant the role of woman as cook in the social organization was more heavily emphasized than the production of offspring.

References

Abi-Bakr, al-Hajj Umar b, *Poverty*. Translated by I.A. Tahir. IASAR/371 (Institute of African Studies, Legon, 1923).

Z. Ali, A.-R. Abizari, Ramadan fasting alters food patterns, dietary diversity and body weight among Ghanaian adolescents. Nutr. J. **17**(1), 75 (11 August 2018). https://doi.org/10.1186/s12937-018-0386-2

J. Allman, J. Parker, *Tongnaab: The History of a West African God* (Indiana University Press, Bloomington, 2005)

J.M. Allman, V.B. Tashjian, *'I Will Not Eat Stone': A Women's History of Colonial Asante* (James Currey, Portsmouth, 2000)

K. Arhin (ed.), *The Papers of George Ekem Ferguson: A Fanti Official of the Government of the Gold Coast, 1890–1897* (Afrika-Studiecentrum, Leiden, 1974)

G. Austin, Mode of production or mode of cultivation: explaining the failure of European cocoa planters in competition with African farmers in Colonial Ghana, in *Cocoa Pioneer Fronts since 1800: The Role of Smallholders, Planters and Merchants*, ed. by W. G. Clarence-Smith, (Palgrave Macmillan UK, London, 1996), pp. 154–175. https://doi.org/10.1007/978-1-349-24901-5_9

A.K. Awedoba, The Peoples of Northern Ghana. National Commission on Culture Website—www.ghanaculture.gov.gh 2006, http://lagim.blogs.brynmawr.edu/files/2015/03/The-Peoples-of-Northern-Ghana.pdf

R.B. Bening, Land policy and administration in Northern Ghana 1898–1976. Trans. Hist. Soc. Ghana **16.2**(1), 227–266 (1995)
R. M. Blench (ed.), *Natural Resource Management in Ghana and Its Socio-Economic Context* (Overseas Development Institute, London, 1999)
R. Blench, *Dagbani Plant Names* (Kay Williamson Educational Foundation, Cambridge, 2012) https://pdfs.semanticscholar.org/6998/8d98a5a51fdf761caa6c47c5dd11fb204919.pdf
T.E. Bowdich, *Mission from Cape Coast Castle to Ashantee* (J. Murray, London, 1819)
J. Brindle, *A Dictionary and Grammatical Outline of Chakali* (Language Science Press, 2017) http://library.oapen.org/handle/20.500.12657/31329
A.W. Cardinall, *The Natives of the Northern Territories of the Gold Coast; Their Customs, Religion and Folklore* (Routledge, London, 1920)
A.W. Cardinall, *The Gold Coast, 1931* (Government Printer, Accra, 1932)
B. Chalfin, *Shea Butter Republic: State Power, Global Markets, and the Making of an Indigenous Commodity* (Routledge, London, 2004)
B. Cruickshank, *Eighteen Years on the Gold Coast of Africa* (London, Hurst and Blackett, London, 1853)
J.M. Dalziel, J. Hutchinson, *The Useful Plants of West Tropical Africa* (Crown Agents for the Colonies, London, 1937)
J. Destombes, Nutrition and chronic deprivation in the West African Savanna: North Eastern Ghana, c. 1930–2000. PhD diss., London School of Economics and Political Science, 2001
J. Destombes, From long-term patterns of seasonal hunger to changing experiences of everyday poverty: Northeastern Ghana c. 1930–2000. J. Afr. Hist. **47**(2), 181–205 (2006)
E.V. Doku, Root crops in Ghana. Ghana J. Sci. **6**, 15–36 (1966)
C.R. Doss, Men's crops? women's crops? the gender patterns of cropping in Ghana. World Dev. **30**(11), 1987–2000 (1 November 2002). https://doi.org/10.1016/S0305-750X(02)00109-2
M.J. Field, *Social Organization of the Gã People* (Crown Agents for the Colonies, Accra, 1940)
C. Fischler, Food, self and identity. Soc. Sci. Inf. **27**(2), 275–292 (1 June 1988). https://doi.org/10.1177/053901888027002005
M. Fortes, S.L. Fortes, Food in the domestic economy of the Tallensi. Africa **9**(2), 237–276 (1936)
R.A. Freeman, *Travels and Life in Ashanti and Jaman* (A. Constable & Co, Westminster, 1898)

A. Gautier, W. Van Neer, The continuous exploitation of wild animal resources in the archaeozoological record of Ghana. J. Afr. Archaeol. **3**(2), 195–212 (25 October 2005). https://doi.org/10.3213/1612-1651-10050

E.N. Goody, *Contexts of Kinship: An Essay in the Family Sociology of the Gonja of Northern Ghana* (Cambridge University Press, 1973)

J. Goody, *Cooking, Cuisine and Class: A Study in Comparative Sociology* (Cambridge University Press, Cambridge, 1982)

J. Goody, E.N. Goody, Food and identities: changing patterns of consumption in Ghana. Camb. Anthropol. **18**(3), 1–14 (1995)

M.D. Graham, W. Report on the vegetable foodstuffs in use at Kintampo. Annual Report. ADM 56/1/458. PRAAD. Gold Coast, N. T., 1902

J.R. Ham, "Who knows tomorrow?": Food Insecurity, distress, and managing the future in Upper West Ghana. University of Georgia, 2016. http://www.secheresse.info/spip.php?article71639

J.R. Ham, Cooking to be modern but eating to be healthy: the role of Dawa-Dawa in contemporary Ghanaian foodways. Food Cult. Soc. **20**(2), 237–256 (2017). https://doi.org/10.1080/15528014.2017.1305827

K.B. Hanrahan, "Mɔn" (to Marry/to Cook): negotiating becoming a wife and woman in the kitchens of a Northern Ghanaian Konkomba community. Gend. Place Cult. **22**(9), 1323–1339 (21 October 2015). https://doi.org/10.1080/0966369X.2014.993360

T.E. Hilton, *Ghana Population Atlas: The Distribution and Density of Population in the Gold Coast and Togoland Under United Kingdom Trusteeship* (T. Nelson, 1960)

A.A. Iliasu, The establishment of British administration in Mamprugu, 1898–1937. Trans. Hist. Soc. Ghana **16**(1), 1–28 (1975)

F. Kröger, B. Meier, S. Tonah, Conflicts and consensu between Migrant Fulani Herdsmen and Mamprusi farmers in Northern Ghana, in *Ghana's North: Research on Culture, Religion, and Politics of Societies in Transition*, ed. by F. Kröger, B. Meier, S. Tonah, (Peter Lang GmbH, Frankfurt am Main, 2003)

D. Kuwabong, Bagre: a Dagaaba celebration of environmental balance between humans and non-humans. J. Dagaare Stud. **4**, 1–14 (2004)

J.D. La Fleur, *Fusion Foodways of Africa's Gold Coast in the Atlantic Era* (Brill, Leiden, 2012)

C. Lentz, Local culture in the national arena: the politics of cultural festivals in Ghana. Afr. Stud. Rev. **44**(3), 47–72 (December 2001). https://doi.org/10.2307/525593

N. Levtzion, *Muslims and Chiefs in West Africa [Electronic Resource]: A Study of Islam in the Middle Volta Basin in the Pre-Colonial Period*, Oxford Studies in African Affairs (Clarendon Press, Oxford; New York, 1968) https://ezproxy-prd.bodleian.ox.ac.uk/login?url=http://hdl.handle.net/2027/heb.02619

A. Logan. A history of food without history: food, trade, and environment in West-Central Ghana in the Second Millennium AD. PhD diss., University of Michigan, 2013

G.B. Louis, *Du Niger au Golfe de Guinée par le pays de Kong et le Mossi, par le capitaine Binger (1887–1889)* (Hachette, 1892) http://archive.org/details/dunigeraugolfed00binggoog

K. Marfo, S. Wiggins, Changing farming practices in the forest Guinea Savannah zone of Ghana, in *Natural Resource Management in Ghana and Its Socio-Economic Context*, ed. by R. M. Blench, (Overseas Development Institute, London, 1999)

J.C. McCann, *Stirring the Pot: A History of African Cuisine* (Ohio University Press, Athens, Ohio, 2009)

J. O'Connell, The withdrawal of the high god in West African religion: an essay in interpretation. Man **62**, 67–69 (1962). https://doi.org/10.2307/2796426

F. Osseo-Asare, "We eat first with our eyes": on Ghanaian cuisine. Gastronomica **2**(1), 49–57 (2002)

F. Osseo-Asare, B. Baëta, *The Ghana Cookbook* (Hippocrene Books, New York, 2015)

M.A. Padmanabhan, The making and unmaking of gendered crops in Northern Ghana. Singapore J. Trop. Geogr. **28**(1), 57–70 (2007). https://doi.org/10.1111/j.1467-9493.2006.00276.x

L. Prussin, *Architecture in Northern Ghana: A Study of Forms and Functions* (University of California Press, Berkeley, 1969)

R.S. Rattray, *The Tribes of the Ashanti Hinterland* (Clarendon Press, Oxford, 1932)

A.B. Stahl, *Making History in Banda: Anthropological Visions of Africa's Past* (Cambridge University Press, 2001)

D. Tait, *The Konkomba of Northern Ghana: Edited From His Published and Unpublished Writings by Jack Goody* (Routledge, 2018)

L. Tauxier, *Le noir de Bondoukou* (E. Leroux, Paris, 1921)

F.K. Tsigbey, R.L. Brandenburg, V.A. Clottey, Peanut production methods in Northern Ghana and some disease perspectives. World Geogr. Peanut Knowl. Base Website **9**, 33–38 (2003)

E.V.W. Vercruijsse, L.M. Vercruijsse-Dopheide, K.J.A. Boakye, *Composition of Households in Some Fante Communities: A Study of the Framework of Social Integration* (Institute of Social Studies, Cape Coast, 1972)

A. Wardell, N. Fold, Globalisations in a nutshell: historical perspectives on the changing governance of the Shea Commodity chain in Northern Ghana. Int. J. Common. 7(2), 367–405 (29 August 2013). https://doi.org/10.18352/ijc.361

M. Warner, R. Al-Hassan, J. Kydd, A review of changes to farming systems of Northern Ghana (1957–94), in *Natural Resource Management in Ghana and Its Socio-Economic Context*, ed. by R. M. Blench, (Overseas Development Institute, London, 1999)

I. Wilks, The growth of Islamic learning in Ghana. J. Hist. Soc. Nigeria **2**(4), 409–417 (1963)

7

Colonialism and Local Foodways

The modern nation of Ghana was conceived in a food fight. In January 1948, the Gã chief, Nii Kwabena Bonne III, organised a campaign to boycott European alimentary and textile goods to protest the high cost of living on the Gold Coast. Among the disputed items boycotted by the residents of the Gold Coast colony were tinned meat and wheat flour biscuits. Chiefs of the different, mostly coastal towns pledged their support and involvement. The boycotters were guided by the slogan "We cannot buy; your prices are too high. If you don't cut down your prices then close down your stores; and take away your goods to your own country."[1] On 28 February, the last day of the boycott, Gold Coast ex-servicemen marched from central Accra to the seat of the British government at Christiansborg Castle to present the governor with a petition. They were fired upon by police, leaving several leaders of the group dead. This in turn led to rioting and became the catalyst for the 1949–1951 campaign for independence.

[1] 'Ghanaians boycott European goods, 1948', Global Nonviolent Action Database, accessed August 26, 2014, http://nvdatabase.swarthmore.edu/.

B. Simpson Miller, *Food and Identity in Nineteenth and Twentieth Century Ghana*,
Food and Identity in a Globalising World,
https://doi.org/10.1007/978-3-030-88403-1_7

Kwame Nkrumah used this event to break away from his role as the General Secretary of the existing nationalist party, the United Gold Coast Convention (UGCC). Composed of merchants, educated elites, chiefs, and prominent farmers, the UGCC had a conservative agenda. Nkrumah and members of the Convention People's Party (CPP), composed of young people he successfully mobilised, began to campaign for Ghana's immediate independence from British rule, and rapidly became the leading nationalist party. The 28 February incident that claimed the lives of former servicemen was considered "the straw that broke the camel's back". It marked the beginning of the process of independence for the Gold Coast, the new nation state of Ghana, the first Sub-Saharan African colony to gain independence.[2] It also marked the beginning of the long struggle for Ghana to become independent of its reliance on imported staple foods, one of the issues that initiated the unrest in 1948.[3]

Irrespective of what the boycott suggested about dependency on expensive imported goods, the main argument of this chapter is that foodways on the Gold Coast were not profoundly transformed by the colonial experience. The tendency to vernacularise imported foods (the translation and/or absorption of a food into a regional repertoire), the reinforcement of local foodways due to the vagaries of monoculture export (cocoa and palm oil) and the resultant food insecurity have contributed to the maintenance of local foodways. Indeed, the retention and evolution of seasonal, varied, abundant, and tasty Ghanaian food and cooking challenges contemporary Western ideas of West African culinary periodisation into pre-colonial, colonial, and post-colonial eras.

Notwithstanding this argument, asking the question of how imported foods were integrated into the main meal of the day is worthwhile and holds significance for understanding the political and cultural history of the Gold Coast and later, Ghana. The purpose of this chapter is to ascertain the impact of twentieth-century nutritional policies on the foodways of Ghana from the colonial period to the present day. Although imports of foods such as wheat flour, sugar, tinned fish, meat and tomatoes rose

[2] Dennis Austin, *Politics in Ghana, 1946–1960* (New York: Oxford University Press, 1970), 72–75.

[3] Osu Mantse Alata and Oyokohene, '14 February 1948 Boycott Speech', ARG 1/1/254, PRAAD, Accra.

steadily over this period, regional cuisines remained a persistent feature of the local landscape. The pattern of consumption whereby new foods were adopted and adapted to local tastes—a feature of Ghanaian foodways that had been in place for centuries—continued.[4] While the colonial and the independent governments of Ghana sought to reduce the desire for and consumption of foreign foods, locals incorporated imported foods into their traditional regional dishes in a manner that preserved the form of the proper meal. This process has not gone without comment by traditionalists.[5] However, the thoughtful mixture of imported convenience foods into the compositions of regional specialties has preserved the structure of the proper meal while simultaneously boosting the consumption of foreign foods.

This chapter considers the effects of colonial nutritional policy on the regional repertoire, reviews the actual ways in which tinned foods, starches, ready seasonings and sugar were utilised in cooking and eating, and considers ideas about the imagined national community of Ghana through the lens of food. This chapter is organised into a brief review of the scholarship of Ghanaian foodways from the mid to late twentieth centuries, and an examination of government efforts to offset the costs of imported foods throughout the twentieth and early twenty-first centuries. The chapter concludes with a review of the relationship between imported food and the national identity of Ghana.

In the late 1990s to the early 2000s, the historical scholarship on foodways began to develop around investigations of national identity and authenticity with work by scholars such as Richard Wilk and Jeffrey Pilcher.[6] Jack and Esther Goody spoke to issues of identity and consumption in mid-1990s northern Ghana directly, and in West Africa more generally. They stressed three forces that could effect change upon a cuisine: famine or shortage, new crops or varieties, or hierarchical emulation.[7] Over the thirty years in which they studied Ghana, they concluded

[4] A. G. Hopkins, *An Economic History of West Africa* (London: Routledge, 1973), 124–57.

[5] Ayensu, *The Art*; Asantewaa Tweedie, *Melting Pot: Ghanaian Cooking with a Twist* (Accra: Asantewaa Publication, 2011).

[6] Pilcher, *Que Vivan*; Richard Wilk, *Home Cooking*.

[7] Jack Goody and Esther N. Goody, 'Food and Identities: Changing Patterns of Consumption in Ghana', *Cambridge Anthropology* 18, no. 3 (1995): 2.

that there had been no change in staple foods or in the methods of preparation, but that increased travel, office work, and education locally and abroad exposed people to different foods. Unchanged were the regional identities as expressed through cooking and eating. What had changed was the greater value placed on cultural diglossia by elites, which is on display in hotels, restaurants, and grocery stores that sell foreign food and regional staples. My work does not seek to explain many of the economic forces behind the growth of imports as the Goodys work does, but rather focuses upon how people slotted imported foods into their existing food systems. As the Goodys asserted, local food may be "kept in the background", but I assert that they still structured daily and seasonal rhythms, and that local food remains at the foundation of Ghanaian identities.

Paul Nugent has also examined ideas about the African nation from the angle of food. Nugent noted that there may have been a convergence around New World starches like maize in West Africa, but that Ghana had not been caught up in the drive to define a national identity with food.[8] Ghana retained and enjoyed its regional foodways, taking pleasure in the variety that came with travel from region to region. Nugent speculated that unlike nations like Senegal and Egypt, Ghanaians did not place a high value upon commensality and that this may be why they had not imagined their nation through food, but preferred regional forms of food consumption. Tuomainen's work on commensality among Ghanaian migrants in London supported Nugent's observations, arguing that kinship structures and the sharing of regional foods enabled the unity of family and community rather than the European concept of commensality.[9] More recent work by Miescher, Sill, and Williams-Forson corroborated the tendency toward regionalism in Ghanaian cuisine.[10] Nugent further observed that Ghanaians transmuted the esteem for the products of their home regions into an internalised definition of national distinctiveness.[11] This chapter is an extension of Nugent's work in that it more closely considers attitudes about food and its relationship to

---

[8] Paul Nugent, 'Do Nations Have Stomachs? Food, Drink and Imagined Community in Africa', *Africa Spectrum* 45, no. 3 (28 March 2011): 87–113., 105.

[9] Tuomainen, 'Eating alone', 35–36.

[10] Miescher, *Making Men,* 79–80; Sill, *Encounters, 294*; Williams-Forson, 'I Haven't Eaten', 75.

[11] Nugent, 'Do Nations Have Stomachs?', 109.

nationalism in Ghana and compares these attitudes to the outcomes of the Ghanaian National Cultural Policy from the 1950s to the present day.

Jonathan Robins too observed that imported provisions were absorbed into the local foodways in Ghana.[12] He examined the development of the colonial nutritional policy from 1900 to 1950. His findings indicated that it was not until the First World War and the Depression that followed those concerns over the costs of imported food were translated into legislative action. Through the late 1940s, a policy of import substitution combined with the modernisation of local food through women's education prevailed. Through a thirty-year period, local methods of food preparation remained unimpeachable even though the concurrent consumption of imports continued to grow. By independence in 1957, wheat, fish, meat, milk, and sugar imports soared. This chapter looks more closely at regulations that affected both local and imported foods, examining the ways in which the adoption and adaptation of foreign foods thwarted governments' efforts to reduce consumption of them.

Jeremy Rich's study of Libreville from 1840 to 1960 argued that in Gabon, as in Ghana, staple starches (manioc and plantain in this case) remained part of the offerings, and that a process of appropriation and borrowing from European imported foods took place. Similar issues faced by some Ghanaian cities were faced by Libreville residents. Intra-ecological transport difficulties, additional provisioning demands of colonial export industries, wars, and ecological changes all factored into the availability of foodstuffs. Unlike Ghana, the social meanings associated with food production in Libreville were closely tied with the work of slaves and women exclusively. This was due in large part to existing social structures that combined with the late-nineteenth-century arrival of slavery there. The meaning of imported foods "became part and parcel of an exchange system where goods embodied reciprocity and respect for African workers."[13] The findings in this chapter largely contradict Rich's findings in Libreville. Local foods and the role they played in social replication worked against the complete elevation of imported foods over local foodways. This chapter also addresses the often-overlooked historical

[12] Robins, 'Food Comes First', 178.

[13] Rich, *A Workman*, 86.

relationship between the feeding of colonies and the wider implications for understanding the maintenance of social relationships in West African nations like Ghana.

Colonial Food Policies

Most colonial governments share a common predicament in that their sole source of revenue is the collection of taxes and the customs receipts on imported goods. This meant that Ghana in the late nineteenth to the early twentieth century relied on palm oil, mining, and cocoa for revenue. Cocoa earnings accounted for 84 percent of the colony's total revenues by 1927.[14] This had the effect of concentrating all the economic power at Takoradi port, and on the administration's ability to tax incoming and outgoing goods.[15] This revenue stream was particularly susceptible to the rising and falling prices of its main export commodities, which could unexpectedly deprive its government of income. After the First World War, food shortages became an increasing problem in the colonies. Colonial extraordinary gazettes, which provided official notification of action taken by the authorities, were issued to control food prices in some towns in Ghana. By 1919, every district monitored prices of local and imported foodstuffs.[16] District Commissioners dictated the prices of local produce, while imported foods could find their own levels.[17] The cost of imported goods soared and was widely protested. This created resentment among local people who felt that the colonial government was intervening on behalf of expatriate merchants at the cost of locals. The

---

[14] Shashi Kolavalli and Marcella Vigneri, 'Cocoa in Ghana: Shaping the Success of an Economy', in *Yes, Africa Can: Success Stories from a Dynamic Continent*, eds. Punam Chuhan-Pole and Manka Angwafo (Washington, D.C: World Bank Group Publications, 2011), 201–17.

[15] Frederick Cooper, *Africa since 1940: The Past of the Present* (Cambridge: Cambridge University Press, 2002), 156, 199.

[16] 'Northern Territories Annual Report with Descriptions of Local Produce and Revenue' (Accra, Ghana: PRAAD, Accra, 1919), ARG 1/17/24; 'Brong Ahafo' (Accra, Ghana: PRAAD, Accra, 1919), ARG 1/17/24; Dieterle, 'Basel Mission Report Aburi'.

[17] Kwabena O. Akurang-Parry, '"Untold Difficulties:" The Indigenous Press and the Economic Effects of the First World War on Africans in the Gold Coast, 1914–1918', *African Economic History*, no. 34 (2006): 58.

coastal areas, where acculturation of European imports was most prevalent, were hardest hit.

For many in Ghana, this prompted a shift to localised social and economic wellbeing. As exhibited by the writings of students in Lady Clifford's *Our Days on the Gold Coast*, in the section entitled "The Lessons of the Letters", students from the coastal region, central forested area, and Lomé (the present-day capital of Togo) wrote letters to soldiers of the British Empire about their future aspirations. Although many boys expressed wishes to be soldiers, engineers, teachers, and accountants, most articulated the idea that being a farmer was the best way to serve their country. In an echo of the concern Reindorf expressed about the cost of provisions in 1895, some boys stated their desire to be farmers stemmed from the lack of availability of food, due to global conflict: "The European provisions on which we rely and boast of, we cannot receive them in large numbers as before."[18] The attitude of many Ghanaians at this time was to solve the problem of soaring imported food prices by increasing production, and in the process, gaining a measure of independence for themselves.

It was the social organisation of Ghanaian society that encouraged local cuisine to remain a prominent feature of the food landscape, as attitudes toward food production remained positive.[19] The local traditional social structure considered food production to be an honourable occupation and an act of service. If we compare Rich's case study of Libreville to Ghana, it becomes apparent that the preference for imported food in Libreville was partially structured by gender and class roles and by local ecology to produce a predominantly consumer society. The late introduction of the slave trade in the mid-nineteenth century (due to its gradual eradication in places like Ghana) slotted into existing local production methods in such a way that was to see the Mpongwe people there rely almost exclusively on slave and female labour for food production: "Culinary practices rested on the foundation of slave labour and social stratification."[20] A challenging topography—the hinterland of the Gabon

[18] Lady Clifford, ed., *Our Days on the Gold Coast* (London: John Murray, 1919), 199.
[19] Robins, 'Food Comes First', 178.
[20] Rich, *A Workman*, 27.

Estuary was characterized by swampy marshland, hills, and ravines—made it easier to import foods from neighbouring countries and from Europe via the ocean. The high price of imported foods came to mean that compensation for waged workers, the majority of which were male, included imported provisions. This resulted in a social organisation whereby the expression of masculinity was tied with leisure and with the eschewing of farming as a respectable occupation.

Much like what was being experienced in Gabon, distribution problems and volatile prices continued to plague the Gold Coast colonial government, compromising their ability to match both military and civilian demand. Despite these difficulties in keeping up with demand, the variety of imported foods continued to grow throughout the early twentieth century. Imported items included rice transported from India and Sierra Leone; tinned butter from South Africa; milk powder and wheat flour from Canada; cabbages and carrots from Nigeria; tinned beef from Australia; tinned butter from Argentina, and fresh meat from the UK.[21] Attempts to substitute tinned fish and imported rice with the local equivalent met with little success. Despite the establishment of rice growing and processing in Ejura, and the study of the fish processing and preservation industry, these efforts were not effective in meeting ever-increasing demand in growing urban centres.[22]

Robins' examination of colonial food policy revealed that by the end of the Great Depression, concerns over plummeting cocoa prices and negative trade balances once again brought attention to rising food prices in the colonies of the United Kingdom. The new science of nutrition was

[21] John Rodger, Governor, 'Memorandum on Sierra Leone Rice', 16 December 1909, ADM 11/1/89, PRAAD Accra; Mr. Packham Gold Coast Marketing Officer, 'Consignment of Cabbage Received from Dept. of Agriculture, Nigeria', 10 March 1941, CSO 8/4/8, PRAAD Accra; Wagon, 'Butter', 1939, CSO 8/8/1, PRAAD Accra; Wagon, 'Telegram Referring to Milk and Wheat Supplies from Canada', 6 February 1942, CSO 8/1/233, PRAAD Accra; Director of Supplies Accra, 'Tinned Beef Order from Australia', 7 April 1942, CSO 8/1/230, PRAAD Accra; Comptroller of Customs, 'Memo on Meat Imports from the UK', 25 March 1941, CSO 8/1/229, PRAAD Accra; Secretary of State for the Colonies, 'Telegram on Argentine Tinned Butter Supplies', 18 June 1945, CSO 8/8/2, PRAAD Accra.

[22] 'Papers of Sir Charles Henry Harper', 1 June 1920 Diary, accessed 25 April 2018, http://www.bodley.ox.ac.uk/dept/scwmss/wmss/online/blcas/harper-ch.html; 'Fish Curing Development', 1921, ADM 23/1/401, PRAAD, Accra; 'Fishing Industry in the Gold Coast By Messrs. John Netherwood And Co., London', 18 April 1933, CSO 8/21/2, PRAAD, Accra.

deployed to decry the health benefits of local foods, which were said to suffer from a lack of variety, contributing to disease.[23] The goal of the colonial government was for the colonies to become self-sufficient in food production, and to discourage the consumption of semi-luxuries; "Things which are not absolutely necessary but which people have got so accustomed to that they are reluctant to do without them."[24] This goal was marketed to the colonies as an initiative to improve their overall nutritional standards by, among other things, increasing their consumption of better kinds of local foods, such as more fruit and pulses. Part of the initiative included the proliferation of more colonial institutions, like schools and agricultural extension programmes. Additionally, women were considered the culprits of poor nutrition due to their use of inferior culinary equipment and inferior food. Local staple foods were deemed too starchy and lacking in nutrition. The solution was to train women to learn domestic science, to use machine-ground starches to replace arduous methods of starch preparation, and for men to take over chores traditionally assigned to women, such as fetching water and firewood.[25] Although women adopted the use of power mills in some cases, social roles with regard to food-processing chores changed little.

The social changes experienced in Ghana in the early twentieth century worked themselves out in such a way that preferences for local foods were maintained. In some cities and centres of commodity and mineral extraction, the colonial government did have trouble provisioning workers and keeping their food costs down. However, the situation was not as dire as that of Libreville, whose residents were so isolated from the hinterland that often the only way in was via boat or aeroplane. Because food production was not solely limited to slaves and women in Ghana, it became a viable way to look after social and economic wellbeing that was not just relegated to an underclass. From the early-twentieth-century accounts of student competitions to plant the most productive plots of cassava and corn, to the urban farming practices of city residents,

[23] Robins, 'Food Comes First', 175.

[24] 'Colonial Economic Policy. (Hansard, 17 December 1940)', accessed 17 March 2020, https://api.parliament.uk/historic-hansard/lords/1940/dec/17/colonial-economic-policy.

[25] F.M. Purcell, 'Nutrition in The Colonial Empire', *British Medical Journal* 2, no. 4100 (1939): 294–96.

including police officers, civil servants, teachers, drivers, and mechanics, farming was widely practiced across society.[26] Urban farming was performed on a part time basis as a supplement to income and household food purchases, and as a hedge against the unpredictable job market in the formal sector. Those who practiced urban farming usually had some exposure to farming in the past and valued food production as a cultural practice (Fig. 7.1).[27]

During the Second World War, despite various methods used to control local food prices, such as the implementation of import substitution schemes, an educational emphasis on nutrition for women and girls, and the introduction of school feeding programmes, the colonial government saw the costs associated with imported foods rising yet again. As in the

Fig. 7.1 Urban maize farming in Accra

[26] Raymond Asomani-Boateng, 'Urban Cultivation in Accra: An Examination of the Nature, Practices, Problems, Potentials and Urban Planning Implications', *Habitat International* 26, no. 4 (December 2002): 597; Miescher, *Making Men*, 67.

[27] Prince Asafu-Adjaye, 'The Tendency to Urban-Farm in Accra: A Cultural Lag-Labor Surplus Nexus', *Journal of Third World Studies* 29, no. 2 (Fall 2012): 177.

First World War, global shortages translated into heavy regulation of local foodstuffs to keep prices stable and to ensure vital enterprises like mining and rubber production were supplied with food. In 1939, Ashanti District Commissioner A. T. Knight drafted an order that severely restricted the intraregional transport of local food: "no plantain, coco-yams, cassava or corn shall be railed from Konongo or Juase railway stations except by the written consent of the District Commissioner, Ashanti-Akim."[28] Konongo was a gold and manganese mining community located in the central forested zone. Regulations were so stringent that the amount of staple starches workers needed was closely calculated: "I hereby authorise Miss Comfort Christian Atta of Pra River to rail once a week one bag of plantain and one bag of coco-yams to Kwahu Prasu for the consumption of Rubber Gangs working in the area."[29] This heavy regulation of the movement of local staples depressed intra-ecological trade and served to reinforce the importance of local food production in each region.

In 1941, the Director of Supplies, Captain A. C. Duncan Johnstone, held a meeting to discuss the control of imported food prices, which were being adversely affected by the high cost of transport. To facilitate control over the price of imports, the colonial government required foreign traders to declare their wholesale stocks of "controlled commodities". Stocks included items such as flour, rice, salt, sugar, milk, lard, tinned fish, tinned meat, tea, tinned butter, and tinned cheese. The cost of transportation was not assumed by the government but passed along to trading firms. Firms reacted with a reluctance to operate in outstations far from the coast. Administrators quickly realised that if prices for foods—which were published in a monthly gazette—varied from, say, Kumasi to the colonial capital of Accra, people would travel to Accra to purchase items at a lower price and resell them in Kumasi.[30] These distribution issues presented a constant threat to the preservation of the stable prices that

[28] 'Draft Order to Restrict Railing of Local Staples from Konogo or Juaso Railway', 1939, ARG 6/12/7, PRAAD, Kumasi.

[29] 'Permission to Rail/Export Foodstuffs from Juaso', 17 September 1945, ARG 6/12/7, PRAAD, Kumasi.

[30] 'Minutes of the Meeting Held in the Director of Supply's Office', 25 October 1941, ARG 1/1/67, PRAAD, Accra.

the colonial government was attempting to establish for imported provisions.

Britain was one country of the 44 nations that participated in what was to become the UN Food and Agriculture Organisation (FAO) just after the end of the Second World War.[31] For the British colonies, this new scheme was in many ways a continuation of the earlier 1936 Colonial Development programme that relied upon many of the same policies, including a focus on the nutritional education of women and girls. In 1944, the Labour Party adopted many of the recommendations of the FAO. Dietary improvements were suggested for vulnerable groups, of which African colonies were deemed to be a part. Recommendations to remedy malnutrition and disease were proposed and short-term and long-term changes in production were implemented. This translated into increased consumption of certain foods like nuts and milk, development of fisheries, and bituminisation of vegetable oils and margarines in all the colonial territories, including Ghana. Additionally, the instruction of girls in cooking methods and the establishment of local Nutrition Committees were deemed of critical importance to the nutritional health of the colony, the reasoning being that woman, as mothers, had the most influence over what the family ate. The waiver of duties on imported foods deemed to be nutritionally superior was also implemented to improve nutrition.[32] In a time of global food crisis, it could be argued that this concern reflected an anxiety about the potential for unrest in the colonies, initiated by high food prices and shortages. Ironically, the prescribed relief included more nutritionally improved foods that required industrial processing and that were imported, thus aggravating the import situation.

The colonial government implemented more school feeding programmes to improve the nutrition of children in the colony. Female pupils were commissioned to meet the demand for labour in the provisioning of nutritional meals in the proliferating schools and hospitals of

[31] 'Documents Relating to the First Session of the Food and Agriculture Conference of the United Nations, Quebec, Canada', 16 November 1945, MSS. Brit. Emp. S. 332 1944–1948, Weston Library, Oxford.

[32] 'Recommendations with Regard to Nutrition in the Colonial Territories', March 1944, MSS. Brit. Emp. S. 332 1944–1948, Papers of Arthur Creech Jones, Weston Library, Oxford.

the Gold Coast. The Central Advisory Committee on Education upheld the Manya Krobo State, located in the present-day Eastern Region of Ghana, as an example of a model school feeding programme. Children paid one penny for a breakfast of kenkey, groundnut, plantain, and stew. Labour was to be supplied by schoolgirls and boys who would work as cooks in school kitchens and farmers on Native Authorities' land. It was recommended that "peppers, tomatoes, and onions are put into every stew" to boost the vitamin content of traditional foods. Accordingly, the production of these non-native vegetables was encouraged on farms.

There was some concern over the demands that cooking would put on the education of female pupils: "There is no objection to schoolgirls assisting with the preparation of meals, provided care is taken that their work does not become a burden to them and prejudices their course in domestic science. There is always a danger that schoolgirls may become cooks' assistants instead of pupils undergoing a systematic domestic science course."[33] Despite this concern for their female pupils, the overall thinking was that working as cooks and being trained in hygienic methods of food preparation would translate into improved nutrition and food safety for the entire colony.

Missionary schools assisted the colonial government with their school feeding initiative by providing data on school-children's eating habits. One problem identified in the institutional school feeding initiative was the custom of children purchasing their midday meals in the market or walking miles home to eat a meal before returning to school in the afternoon. This situation was reminiscent of the eating arrangements made in missionary schools in the nineteenth century whereby students and their families arranged their own meals. It was decided that students were not getting the nutrition they needed in the markets or at home as those meals were too heavy on starch and deficient in protein. K. A. Busia, Assistant District Commissioner of Sekondi/Takoradi (and future president of Ghana from 1969 to1972) was employed to study the domestic economy of the people in his district. He noted the business practices of women who supplied students with meals, stating that a "lady" had rented a space from the Sekondi Social Centre where she provided up to

[33] 'Meals for Schoolchildren', 1946, ADM 23/1/2768, PRAAD, Accra.

15 children with a two pence meal of starches that included kenkey, rice, cassava, gari, and fufu served with a stew.[34] He thought it an experiment worthy of replicating for the School Feeding Programme. Irrespective of the sponsorship of the state in setting up a program whereby students were fed in school dining rooms, enterprising women made it their business to offer to feed students just off school property, a practice that continues to the present day. Commercial food vendors can be seen plying their trade at the Legon University all the way down to the primary school level.[35] However, the District Commissioner at the time wanted the entire operation standardised, specifying the ratio of cooks to children, as well as the dimensions of the kitchens and the types of equipment to be used. The nutrients of fish and fruits, as well as that of imported tinned meats and yeast, were heavily promoted.[36] Standardised initiatives like the school feeding programme had the effect of increasing the consumption of imported goods in institutions due to nutritional requirements which were met with enriched, imported foods.

Other opportunities to cook in institutions such as those in the Central Hospital Canteen in Kumasi, were coveted and sought after by local women. One Rebecca Korley sent a letter of application in 1957 to pay monthly rent to run the canteen using her own utensils and stewards, in what she assured the administrator would be with "hygienic cleanliness".[37] She was turned down in favour of the reintroduction of the Red Cross Society, which had run a canteen until 1952. The reason for the reopening of the canteen was in the results of a survey in 1957 conducted by Staff Welfare officer Alfred Moses Dunkor. Dunkor learned that many of the unmarried male nurses had difficulty obtaining good food. Some of the female nurses had no maid servants to do their cooking for them and had trouble getting meals. Additionally, it was difficult for staff to make

---

[34] Dr. K.A. Busia, 'Sekondi-Takoradi Social Survey—School Feeding', 9 December 1947, ADM 23/1/2768, PRAAD, Accra.

[35] Sara Stevano and Deborah Johnston, 'Childrens Food Consumption Away From Home and Food Provisioning in Accra' (SOAS Food Studies Centre, SOAS, University of London, 27 January 2017).

[36] Busia, 'Sekondi Social Survey', 1947.

[37] 'Application to Run Canteen at Central Hospital, Kumasi', 22 January 1957, ARG 13/1/11, PRAAD Kumasi.

the trip to town between midday and one o'clock in the afternoon as the break was too short to purchase and eat lunch. The new staff canteen was to be open from eight o'clock in the morning to two o'clock in the afternoon and was to serve tea, coffee, cocoa, Ovaltine, milk, sugar, butter, cheese, marmalade, biscuits, Marmite, sardines, corned beef, cakes, and minerals. There were some local foods included such as rice, yams, and palm nut soup.[38] Notably, much of the food on offer was imported and its primary ingredient was often imported sugar.

Food Security and Political Stability

As previously stated, during and after the colonial period, the main source of government revenue was derived from duties on goods that entered and left the ports. This did not provide the basis of a strong national economy for Ghana and fostered regional tensions. The existing economic base and transportation systems meant that long-term food security issues were prioritised. During the transitional period when Nkrumah was made Prime Minister, he became the architect of the National Food Board. Inaugurated in January 1953, its purpose was to advise the government on "the construction of feeder roads to facilitate the provision of supplies for labour engaged on the construction of Tema Harbour and the proposed Volta River Project, on guaranteed prices, food reserves and other matters".[39] The National Food Board represented an attempt to resolve inherited food distribution issues and demonstrated an appreciation for the impact food security had on Ghana's ability to transition from an export economy, to one that was diversified and included manufacturing and services.

These attempts to address food security issues were met with little success and were seized upon by those opposed to Nkrumah and the CPP's handling of the issue. In 1954, Nkrumah and the CPP were facing considerable tensions surrounding farm productivity and food security. On

[38] 'The Kumasi Hospital Canteen', 13 June 1952, ARG 13/1/11, PRAAD, Kumasi.

[39] Kwame Nkrumah, 'Circular from Prime Minister Kwame Nkrumah to Chief Regional Officers on Behalf of The National Food Board', 1953, ARG 2/8/27, PRAAD, Accra.

12 May 1954, Ghana's only daily newspaper at the time, *The Daily Echo,* published a commentary entitled "Grow More Flags" which included:

> The C.P.P. government's campaign for the growing of more food failed and Government reports show no concrete increase in food production during the three years of Nkrumah's reign. The Housewife found things gone worse. Hence, the average reflective person can rightly term the present hoisting of C.P.P. flags as the Government's understanding of grow more food.[40]

Despite the criticism, Nkrumah pressed on with programmes that prioritised food security. He became prime minister upon the full independence of the Gold Coast in March 1957, becoming president when Ghana later became a republic in 1960.

The 1962 Programme for Work and Happiness and the 1964 Seven Year Development Plan were designed to eradicate the colonial export structure by increasing the internal production of foodstuffs and consumer goods. Policies included agricultural diversification and import substitution. Nkrumah's plan for Ghana was to develop a self-sustaining economy balanced between industry and agriculture, "providing a sufficiency of food for the people and supporting secondary industries based on the products of our agriculture".[41] Agriculture was diversified and mechanised wherever possible, and state farms produced rubber, palm oil, banana, citrus fruits, cereals, and vegetables.

Nkrumah's agricultural policies proved to be disastrous. Starting as one of the most prosperous African economies at the time of independence, by 1965 Ghana was virtually bankrupt. Independence came amid an economic boom, when the international market value of cash crops such as cocoa, minerals, and groundnuts was high. Rainfall was also good during those years. West African economies grew between 4 percent and 6 percent per year.[42] However, Ghana was to experience an unanticipated fall in the price of cocoa:

[40] 'Grow More Flags', *The Daily Echo*, 12 May 1954, 12.

[41] June Milne, *Kwame Nkrumah, a Biography* (London: Panaf Books, 2006), 107.

[42] Martin Meredith, *The State of Africa: A History of the Continent Since Independence. Martin Meredith* (London: Simon & Schuster, 2013), 41.

> The world price for cocoa, the backbone of the country's economy at the time, either due to international manipulation to subvert the economic programs of the Nkrumah government or due to an inter-play of market forces, plummeted so sharply that the country needed a large dose of external support to maintain financial solvency. By 1960, a ton of cocoa beans on the London Exchange was estimated at 240 pounds on average; by August 1965, it dropped to an unprecedented low of 91 pounds.[43]

In addition to the strain created by the crash of the price of cocoa, between 1959 and 1964 Ghana spent 430 million pounds on development projects, financed mostly by the Cocoa Marketing Board, a government-controlled body. In place of the prosperity experienced just after independence in 1957, food shortages, higher taxes, and foreign debt became the norm in the late 1960s.[44] Furthermore, a shortage of food supplies to the cities made it more profitable for farmers to switch from cocoa to food crops such as maize, millet and yams.[45] Chronic food insecurity in urban areas and a retrenchment of local food stuffs in rural areas became the norm. Although Nkrumah's failed policies were blamed for food shortages, structural food distribution problems and macroeconomic export crop issues played a part in the retrenchment of local foodways and served to reinforce regional food production and cooking methods.

Nkrumah and the CPP were overthrown in a coup in 1966, which was followed by three years of military rule, a brief return to civilian rule under President Busia in 1969–1972, and then another coup and military regime led by General Ignatius Acheampong. The post-Nkrumah era however, was a continuation of economic policy patterns established during the initial post-colonial time period. In 1972, General Ignatius Acheampong came to power under the banner of the National Redemption Council (NCR). Seeking to stabilise the prices of imported staples, Acheampong banned the direct import of essential goods like corned

[43] Boni Yao Gebe, 'Ghana's Foreign Policy at Independence and Implications for the 1966 Coup D'état', *Journal of Pan African Studies* 2, no. 3 (2008): 28.

[44] Meredith, *The State of Africa*, 186.

[45] Lynne Brydon and Karen Legge, *Adjusting Society: The World Bank, the IMF, and Ghana* (London: I.B. Tauris, 1996), 12.

beef, sardines, milk, and sugar by foreign companies. A subsidy of 23 million Ghanaian cedis was then provided for the government to purchase these same essentials to stabilise the volatile prices of these imports.[46] Operation Feed Yourself strove to fill the gaps created by the restriction of imports by promoting backyard farming in the cities and supporting farmers. Generous loans, advice on irrigation projects, and subsidies of seeds, fertiliser, and pesticides were provided to farmers. Students were enlisted as volunteers to harvest sugar cane to boost national production. This project was aided by rising world market prices, and two years of good rains.[47] In an echo of Nkruma's policies, this project reflected the desire for Ghana to be self-sufficient in its food production.

The Acheampong administration kept some farmers focused on local food production but failed to leverage the rising world market price of cocoa by passing on the rise in price to cocoa farmers. Cocoa farmers responded by smuggling cocoa to neighbouring Togo and the Côte d'Ivoire in exchange for hard currency in the form of CFA (*Communauté Financière d'Afrique)*, which was then used to purchase restricted essential commodities. The policy of sacrificing cocoa farmers profits effectively neutralised government efforts to lessen dependence on imported sugar and tinned meat.[48] While these policies did strengthen the consumption of local foods, they were unable to reduce dependence upon foreign staples.

In 1980 36 Sub-Saharan governments signed on to IMF and World Bank stabilisation programmes which required structural adjustments. Foreign aid became a crucial component of African economies, taking over key functions of the government and becoming a major employer, second only to the state.[49] From 31 December 1981 Ghana was ruled by J. J. Rawlings's austere Provisional National Defence Council (PNDC) military regime. In 1982, Ghana became one of the African governments to ask for aid as it was close to economic collapse. Food supplies and

[46] Gocking, *History of Ghana*, 167.

[47] Ibid., 167.

[48] Ales Bulir, *The Price Incentive to Smuggle and the Cocoa Supply in Ghana, 1950–1996*. (Washington, D.C.: International Monetary Fund, 1998), 14.

[49] France Moore Lappe, Joseph Collins, and David Kinley. 'Aid as Obstacle: Twenty Questions About Our Foreign Aid and the Hungry', *Food Policy*, 6, 3 (1981): 98.

production levels were at an unprecedented low. Many roads were impassable. A severe drought plagued the region and caused one million Ghanaians to be expelled from Nigeria and repatriated. Rawlings went to work embracing the offered World Bank and IMF restructuring programme. He underscored the debt the nation owed to farmers, stating, "We are acknowledging the historic debt of the whole nation to the farmer and have thus repudiated the monstrous injustice of a past in which we virtually ran the machinery of the state on the tired backs of rural producers and provided little for their basic needs."[50] As in the first World War and Nkrumah's rise to power, Rawlings saw the promotion of increased food production as the solution to the pressures of the external restructuring program.

As part of their restructuring policy, the PNDC devalued the Ghanaian cedi (GH₵) currency in 1983. That year the exchange rate started at three to one with the US dollar and plummeted to the rate of 450 cedi to the US dollar by 1992. The goal of devaluation was to make the chief export commodity of cocoa more competitive on the global market by making it less expensive to purchase. This had the unfortunate side effect of raising the cost of the imported staples that Ghanaians had come to rely upon. Over 60,000 public service sector jobs were eliminated, and public sector organisations such as Ashanti Goldfields were privatised. Rawlings experienced mixed results from the implementation of these reforms. Foreign debt, which rose in part due to subsidised purchases of imported staple foods, more than doubled, reaching 3.3 billion dollars from the period of 1983 to 1988.[51]

The revival of the 1950s National Cultural Policy would become one way for Ghana to mitigate the foreign debt of the 1980s. In the early years of Nkrumah's administration, the government had begun to institutionalize restrictions on opposition parties and advanced a system of exhibitions, documentaries, and national art that presented the vision of a homogeneous nation-state.[52] The Interim Committee for the Arts

[50] Meredith, The State of Africa, 372.

[51] Ibid., 372.

[52] Janet Hess, 'Exhibiting Ghana: Display, Documentary, and "National" Art in the Nkrumah Era', *African Studies Review* 44, no. 1 (2001): 63.

Council of the Gold Coast was established by the Nkrumah government in 1955 for this purpose. The 1980s Heritage Tourism Scheme, in contrast to the earlier Arts Council which invested in projecting homogeneity, took a different approach that included a focus upon Ghana's various ethnic groups. It contributed to the conception of a national character that incorporated multiplicity and allowed citizens a role in the crafting of the national identity as their role as entrepreneurs expanded. Local businesspeople ran the arts markets and gave guided tours of local attractions from the James Town lighthouse to the historic Larabanga mosque in the Northern Region. National and ethnic approaches of identification worked in concert to create a distinct nationalism that converged around common traits and local idiosyncrasies. However, local cuisines were conspicuously absent in the marketing efforts.

National Cultural Policy and State Entertainment

The National Cultural policy began with Nkrumah as a mechanism to build national pride and identity. Nkrumah decreed the creation of councils, organisations, and centres to forge a common national identity and culture. His goal was to incorporate disparate artistic traditions into a common Ghanaian identity based upon a uniquely Ghanaian sensibility. He strove to elevate a Ghanaian rural character he believed had been denigrated by the effects of colonialism and missionary work. The Arts Council organised various festivals whose venue changed annually among the regional capitals, thus promoting a sense of nationhood.[53]

The first incarnation of the cultural policy did not mention food of any kind, concentrating instead on visual culture such as cloth or the organisation of Miss Ghana contests to mark Ghana's Independence Day celebrations.[54] The process of negotiating a national identity using visual

[53] Sandra E. Greene, 'Developing the Arts for Development: Perspectives on Ghana', *Africa Notes* (1998): 1–8.

[54] Jerry Orhin Yorke and Emmanuel Kodwo Amissah, 'An Overview of Kwame Nkrumah's Cultural Policies on Ghana's Visual Culture', *Research Journal of Humanities and Cultural Studies* 3, no. 3 (2017): 12.

imagery is precisely the focus of Harcourt Fuller's development of the term *symbolic nationalism*. Fuller defined this term as Nkrumah's accessing of Western traditions and grounding them in localised cultural symbols and philosophies of governance.[55] A prominent example of the use of localised cultural symbols was the application of the image of Tetteh Quarshie to the new Ghanaian currency. Quarshie was a Gã blacksmith who was credited with bringing the first cocoa seeds from the Spanish island of Fernando Po in 1879. His image, and that of a cocoa pod, are featured as a watermark on the Ghanaian one-cedi bill.[56]

Eventually, the cultural policy grew to include foodways. For a time, Ghana was under military control intermittently beginning in 1979. Flight Lieutenant Jerry Rawlings was head of state three times between 1979 and 2000. The first time, he assumed control as the chairman of the Armed Forces Revolutionary Council in 1979. The second time, he was chairman of the PNDC from 1981 to 1992, after which he formed the National Democratic Congress (NDC) and was elected president twice between 1993 and 2000. Rawlings stood down in 2000 due to the constitutional two-term presidential limit implemented in the 1992 Constitutional Referendum. John Agyekum Kufuor succeeded him and served as president from 2001 to 2007 with the New Patriotic Party (NPP). It was not until 2004, during Kufuor's administration, when the cultural policy first included the mention of food. What follows is a section of the 2004 Cultural Policy:

> 10.4 Foods
> 10.4.1 Ghana has a rich diversity of foods and culinary cultures from its diverse ethnic Cultures. The state shall:
> a. actively support research into production and preservation of local foods; and the compilation of traditional recipes and methods of preservation.
> b. Encourage the consumption of Ghanaian cuisine from all parts of the country and discourage the over dependence on imported foods.
> c. Explore the nutritional values of our local foodstuff and promote them.
> d. Encourage the introduction of cuisine from other African Cultures.

[55] Harcourt Fuller, Building the Ghanaian State: Kwame Nkrumah's Symbolic Nationalism (New York: Palgrave Macmillan, 2014).
[56] Ibid., 184.

> 10.4.2 Ghanaian dishes shall be a predominant feature of menus at State functions and in public catering institutions.
> 10.4.3 Ghanaians shall be encouraged to develop a culture of producing what they eat and eating what they produce.[57]

In contrast to the total absence of the mention of foodways in the immediate post-colonial era, here for the first time there was an emphasis on the richness of the regional variation in foods, and a recognition of migratory influences on Ghanaian cuisine. The emphasis was on increased agricultural production to meet the needs of Ghana's citizens for sustenance and economic empowerment, a direct continuation of Nkrumah's economic aims from the early 1950s.

While locally produced food and the stated goal of independence from imports remained the focus of the Cultural Policy, there was still no sign of the conscious development of a national cuisine, but rather an emphasis on the preservation of traditional regional foodways. Fieldwork conducted in 2014 reflected an attitude of pride toward the regionalism of Ghanaian cuisine. Individuals were interviewed about what food they most often consumed, what they thought the national dish was and why, and what role they thought the government should play in promoting local food. Results of this fieldwork yielded several examples of those who stated that food was resolutely regional, feeling it was a source of national strength, making such comments as it brings "a blend of something extra", or it provides "variety, it is healthy, and interesting".[58]

This attitude of tolerance, and indeed of the welcoming of regional foodways, was found in every sector of society. John Kwadwo Osei-Tutu, professor of history at the Norwegian University of Science and Technology and visiting professor at the University of Ghana, stated that:

---

[57] National Commission on Culture, 'The Cultural Policy of Ghana', http://www.ghanaiandiaspora.com/wp/wp-content/uploads/2014/09/Cultural.-Policy-Final.pdf.
[58] Simpson Miller, 'Food and Nationalism', 70.

"Ghana has no need for a national food."[59] As one of the respondents who felt food was regional in Ghana, he asserted that it would be divisive if one food was chosen, and that for a politician to attempt this would be disadvantageous to his or her career. Osei-Tutu emphatically stated, "There will be no national food of Ghana. Not in my lifetime, my children's lifetime, or in my children's, children's lifetime". He went on to say that he felt that the football team and the national anthem were more important to the inspiration of national pride than an identifiable Ghanaian national cuisine.

Some respondents believed that there were more important things for Ghana to focus upon rather than the development of a national cuisine. Ama Akoto, Senior Quality Assurance Officer at the Ghana Tourism Ministry, responded to the query about whether she believed a country needed a recognisable food or cuisine to be considered a great country, with "Food is not everything". She believed that cleaning up the beaches was more important to attracting tourism than catering to the culinary tastes of tourists. She admitted that cuisine is important in as much as the possibility of a tourist getting ill tarnished the reputation of a country.[60] However, she did not believe that it would benefit Ghana's tourism marketing to develop a national dish to sell to tourists.

Instead of concerns over national identity, fieldwork conducted in 2014 indicated a more pressing concern over economic self-sufficiency, as stated in the 2004 Cultural Policy. The overriding issue in most responses about cuisine seemed to reflect a concern for the economy and the generation of jobs more than a concern over the marketing of a national identity with cuisine. Many respondents agreed with Ms Akoto in their assessment of the low importance of a recognisable national cuisine for Ghana. Fully 30 percent of those surveyed believed it was not necessary to have a national cuisine, while 47 percent did. Of those who did respond positively, 21 percent believed it would benefit the tourism industry, and 25 percent believed a recognisable national cuisine would

[59] Dr John Kwadwo Osei-Tutu, Personal interview with Brandi Simpson Miller at University of Legon, History Department, 27 May 2014.

[60] Ms. Ama Akoto (Senior Quality Assurance Officer at the Ghana Tourism Ministry, Accra), in conversation with the author, 28 May 2014.

boost the economy and/or employ more Ghanaians.[61] Foreign foods seemed to be perceived as problematic for the economy, contributing to insecurity due to fluctuating pricing, and representative of neo-colonial domination.[62]

In the 1980s during the Structural Adjustment period, the Ghanaian people were expected to participate in the process of creating the national identity, demonstrating that the national conversation was as much a bottom-up exchange as a top-down phenomenon. In fact, the regional character of Ghanaian food today, first initiated by its ecological situation and sustained by its colonial policies and troubled transport and economic systems, has been embraced by the citizenry as a national good, and as something that makes them unique among nations. Over time, and as reflected in changes in the official national cultural policy, the regional character of foodways in Ghana came to be regarded as a source of national strength. Beginning with a policy that wholly excluded any mention of food, the more recent cultural policies highlight pride in a diverse nation that includes recognition of cuisines from different regions and those of other African nations. The study of cuisine as an aspect of Ghana's material history reflects a national cultural complexity that is a reaction to the forces of the colonial participation in global capitalism. "When we use the term material life, it implies at once not only long-term processes of work and domestic life but also the way those processes are conditioned at higher levels by the market economy and capitalism."[63] The regional quality of foodways in Ghana is both an ecological and cultural inheritance as well as a colonial legacy, and a rejoinder too of the inherited colonial economic and transportation systems that served to reinforce gaps in consumption and regional food boundaries.

Regardless of the stipulations of the 2004 cultural policy, Ghanaian dishes were not a prominent feature of menus at state functions and in public catering institutions. Cuisine served at state functions in state-owned hotels, and at offsite locations such as the State House, usually

[61] Miller, 'Food and Nationalism' ', 79.

[62] Agbeve Worlali (student, Department of Family and Consumer Sciences, University of Ghana, Legon), in conversation with the author, 27 May 2014.

[63] Robert Blair St George, ed., *Material Life in America, 1600–1860* (Boston: Northeastern University Press, 1988), 5.

featured continental fare. The reverse of the formality of native dress in Ghana, there is an association between formality and continental cuisine, while regional cuisine is considered informal. According to David Fordjour, Executive Chef of the state-owned La-Palm Royal Beach Hotel, state banquets can present a logistical problem as "African foods are finger foods and soups." Fordjour admitted that for most catered state functions, continental dishes were offered as serving Ghanaian dishes becomes an operational issue in more formal settings, where it is nearly impossible to individually plate five hundred servings of fufu. Buffet style is possible but is less formal and not always an option depending on what is requested.[64] The designation of continental food as formal reveals a historical value of cosmopolitanism and the willingness to display their worldly knowledge and the ability to speak whatever language their guests prefer in the form of international food.

An argument could also be made for the use of European food at state functions as a unifier across regions, effacing regional, class, and cultural differences. Before the National Assembly in 1959, Nkrumah made it very clear what the priority was for his policies: "Ghana is a unity and in this small country there is no room for regional and tribal groups [,] each emphasising their own differences from the rest of the country at the expense of national unity."[65] As in radio, museums, film, and arts of the time, food at state functions may have served as a mechanism for advancing the unifying policies of the nation, while simultaneously offering a carefully-managed vision of a unified and modern nation-state.[66]

The La-Palm Royal Beach Hotel featured six eateries, two of which showcased Ghanaian food and one, The Royal Dragon, which featured Chinese. The Continental Restaurant at La-Palm highlighted some local Ghanaian dishes, alongside a large selection of international main courses. Spatially, the fare was organised to preference Western and Chinese foods. The La-Palm Royal featured the beachside Ghanaian Village Restaurant, which was billed as a pan-African eatery that almost exclusively featured

---

[64] David Kore Fordjour (Executive Chef, La-Palm Royal Beach Hotel, Accra), in discussion with the author, 3 July 2014.

[65] Kwame Arhin, *The Life and Work of Kwame Nkrumah: Papers of a Symposium Organized by the Institute of African Studies, University of Ghana, Legon* (Trenton: Africa World Press, 1993), 94.

[66] Hess, 'Exhibiting Ghana', 68.

Ghanaian regional foods. Housed on a thatched patio adjacent to the beach and located furthest from the main building, the diner was treated to first-class service with an Atlantic Ocean view. However, the beach was enclosed with gates, due to the propensity for a segment of the Ghanaian population to reside on the beach. The gate protected diners from hawkers and some of the more unsavoury aspects of beach-living near Accra. The spatial organisation of foods at state hotels could reflect tensions between local food promotion and foreign food. It could also signal a desire to make foreigners comfortable. No Ghanaian expects foreigners to know of their local cuisines, but most express pleasure when foreigners convey an interest in and an understanding of local foods (Fig. 7.2).

One of the more successful nationalistic government strategies was that of the National Best Farmer. In a modern rendition of the traditional harvest commemoration, the Ghanaian government initiated the Farmer of the Year celebration, beginning in 1985, to recognise and celebrate its farmers. The 1980s were times of volatility in terms of food security due to lack of rainfall and the revaluation of the currency. Increasing

Fig. 7.2 Labadi Beach English Breakfast Buffet

urbanisation further threatened food security in Ghana. The celebration was conceived to show appreciation for farmers' efforts during times of drought and economic volatility, to encourage youth to choose farming as a profession, and to introduce modern methods and government policies. Since 1988, the first Friday in December has been set aside as a national holiday to honour the most successful farmers at the national level. There were many categories, including cocoa, foodstuffs, animal husbandry, and poultry. A national "best farmer" was then chosen from this category, and the winner was awarded with a car, cash, farming equipment, and a house. Past themes included "Grow More Food" or the 2013 theme, "Reducing Post-Harvest losses for Sustainable Food Security and Nutrition".[67]

Today, more than 70 percent of Ghanaians remain subsistence farmers and Ghana continued to face post-harvest losses of food staples, especially of grains and cassava.[68] These post-harvest losses were mainly due to difficulties in transportation and lack of appropriate storage facilities, a reoccurring problem in Ghana. These events underscored the primacy of food availability in policy and planning for the nation, in that the focus remains on increasing production. Post-harvest losses too have been a concern since the Nkrumah era and are still a challenge today. This national event was also one of the few ways that the Ghanaian government fostered a feeling of national pride through food production instead of focusing upon specific regional dishes.

Chop Bars and Dining Out

Long before colonial nutritional policies were transmuted into school feeding programmes and domestic science courses for women and girls, women in southern Ghana experienced a significant change in status due to the gradual monetisation of the economy. In Asante, as export agriculture and monetisation expanded from the late nineteenth century to the

[67] '2013 National Farmers Day Launched', Official Portal, Government of Ghana, accessed September 8, 2014, http://www.ghana.gov.gh/index.php/2012-02-08-08-32-47/features/1460-2013-national-farmers-day-launched.

[68] '2013 National Farmers Day', 2014.

1940s, women became more insecure as their engagement in the cash economy increased. A crisis in gender relations ensued, whereby more of a woman's labour was dedicated to cash-crop production on her husband's farm rather than labouring on family farms. At the same time, men's rights to their children's labour were enhanced while their contribution to their children's maintenance declined. As women faced declining provision for the household and more demands put upon their labour, many took up trading. The first generation of colonised women were market pioneers who sold essential goods like food, cloth, and alcohol. Women combined cooking, trade and travel to make a place for themselves, becoming central to the internal economic and material development of the colonial export economy.[69]

Much like Asante women during the same time period, Gã women in Accra experienced significant socioeconomic change in status when the existing corporate kinship organisation was significantly affected by global capitalism. Women's position in their patrilineal organisation gave them less access to land, which was becoming privatised due to the impact of the export economy. While men concentrated on gaining land and education, many women abandoned farming and production in favour of marketing, relying upon their families as labourers. Using their traditional rights to individual economic transactions and separate property, they traded in local, imported, and prepared foods.[70]

People have been consuming readymade meals in Ghana since the seventeenth century. Prepared food sold in markets by enterprising Ghanaian women was unique in West Africa, predated colonial rule, and arguably preceded the development of fast food in the West.[71] Regardless of the popularity of the sale of cooked food from the seventeenth century on, this activity did not readily translate into western style restaurants in Ghana until the late nineteenth and early twentieth centuries: "Cooking in the south is still a fundamentally private act with connotations for marriage and/or sexual liaison."[72] Cooking retained a special status in the

---

[69] Allman and Tashjian, *'I Will Not Eat Stone'*, 13–18.

[70] Claire C. Robertson, *Sharing the Same Bowl?: A Socioeconomic History of Women and Class in Accra, Ghana* (Bloomington: Indiana University Press, 1984)., 247.

[71] McCann, *Stirring the Pot*, 129; De Marees, *Description*, 62.

[72] McCann, *Stirring the Pot*, 129.

role of social reproduction, centred as it was on the division of labour within a marriage.

Despite the cultural values about cooking which are extant to this day, the late nineteenth and early twentieth century colonial enterprises began to affect a shift in this cultural practice. The capitalist mode of production and the influence of Christianity and western education all impacted local cultural ideas about social interaction. Ideas of leisure time as distinct from work time solidified during the colonial period, particularly during World War I with the stationing of foreign troops in towns like Sekondi and Accra during the war. These troops needed entertainment and needed to be fed. They contributed European style drinking bars, band music, and comic opera to urban culture in Ghana. Highlife music emerged from this burgeoning cultural background, when dance bands fused Akan dance rhythms and melodies with western instruments and harmonies.[73] Dining outside of the home and the consumption of alcohol developed into new concepts of leisure. These cultural changes combined with capitalist colonial enterprises between 1890–1919 in industries such as palm oil production and gold extraction. Mechanized mining and railway towns, sited in places in which there were no densely inhabited settlements, attracted a heavily male labour force and created a demand for sellers of cooked food and retailers of alcohol.[74]

The demand for the public consumption of traditional cooking in new towns was reflected in the proliferation of restaurants and informal and inexpensive eateries called "chop bars", which sold indigenous heavy foods. Chop bars were operated by women in urban areas, at rest stops, and near markets in rural areas. Customarily patronised by men, today chop bars have thought-provoking names such as "Don't Mind Your Wife Chop Bar" or "God First Chop Bar" that either reflect the owner's values or deride cultural norms.[75] By 1928, there was such a profusion of both chop bars and restaurants in Kumasi that they had begun to attract

[73] Emmanuel K. Akyeampong, *Drink, Power, and Cultural Change* (Portsmouth: Heinemann, 1996), 101.

[74] Ibid., 48.

[75] 'The Wonderful Ghanaian CHOP BAR Experience', *GhanaCelebrities.Com* (blog), 25 September 2014, https://www.ghanacelebrities.com/2014/09/25/wonderful-ghanaian-chop-bar-experience--tried/; Osseo-Asare, *Food Culture*, 35.

the attention of the colonial government. The Inspector General of Police instituted an investigation of chop bars, inquiring about the types of food served, and the manner of presentation and consumption.

The Inspector General inquired of the Police Commissioner whether tinned meats such as corned beef, bologna sausage, and fruit salad on bread, *kaklo* (savory cassava or plantain fritters), or *kenkey* could be considered a meal. The Inspector also inquired about the classification of these informal eateries where labourers sat on the ground to eat their meals served on a banana leaf or out of a communal "native pot" using their fingers. Where they or were they not to be classified as restaurants and to pay the same licensing fee? In reply, the commissioner suggested that there be two classes of restaurant license. Class A was for Europeans and the "Educated African" who ate in European style eateries with tables and cutlery. These types of restaurants required washing up facilities and toilets. Class B eateries would have restricted hours and be slightly cheaper than Class A restaurants licenses, with fewer appurtenances, "but complying with all the other requirements" in terms of sanitation and toilets.[76] The regulation of chop bars and restaurants demonstrated that by the late 1920s eating meals outside the home had become an ubiquitous practice for all classes of working men in towns like Kumasi.

The Great Depression amplified the draw of young men and women to cities looking for work in rapidly expanding urban environments. Sekondi grew from 4,095 to 16,953 in the thirty years from 1901 to 1931, and the population of Accra swelled from 19,582 to 60,726 during the same time period.[77] At the same time that cities were experiencing explosive growth, the sale of cheap imported Dutch gin was prohibited by the colonial government. Women short of money took to retailing *akpeteshie*, a home brew made from distilling palm wine or sugar cane.[78] From this initial marketing of alcohol, women would add the sale of traditional food to expand their businesses.[79] Many chop bars started as family businesses, being inherited from a mother or an aunt. At times,

[76] E. K. W. Thompson, 'Inspector General of Police to Commissioner of Police Ashanti', 1928, ARG 1/1/138, PRAAD Kumasi.

[77] A. W. Cardinall, *The Gold Coast, 1931* (Government Printer: Accra, 1932), 101–2.

[78] Akyeampong, *Drink*, 104.

[79] Robertson, *Sharing the Same Bowl?*, 131.

some women who sold kenkey, cooked yam, or sold drinks in the market decided to expand to add meals to their offerings. Other times, the opening of a chop bar was an act of desperation, especially if a woman's husband deserted her and she found herself having to support children on her own.[80] Many women took advantage of the opportunity to support themselves and their children by catering to working class men from various regions in Ghana and West Africa.

Going to the chop bar was considered a treat for working-class people who could not otherwise afford to dine out.[81] Menus were flexible as it was possible to order soup and fufu with no meat, to pay a fee for the addition of meat, or even to bring your own tinned sardines to place atop your soup and starch.[82] The ability to cater to every budget made a traditional 'meal out' accessible to a great many working class people. Chop bars too were places where one could openly encounter other working men of different age groups, ethnicities, and nationalities. Consequently, chop bars provided a place where working culture began a process of urban creolisation, combining Western social drinking and eating habits and synthesising them with local foods and forms that were reflective of their wage labour experiences.[83]

It's role in leisure time aside, chop bars were generally viewed as suspect by middle class Ghanaians for a variety of reasons: "A man with no wife must of necessity know how to cook or else he would be condemned to a life of chop bars."[84] This statement revealed that eating your main meal at a chop bar every day was not an attractive prospect because chop bars were known for taking shortcuts. Examples of common substitutions included powdered tomatoes and peppers instead of freshly ground ingredients, or the replacement of Maggi seasoning cubes instead of local

[80] Laura Alfers and Ruth Abban, 'Occupational Health & Safety for Indigenous Caterers in Accra, Ghana', *WIEGO* (2011): 24.

[81] Kwame Frimpon, *The Boy in the Oversized Smock: School Memories in Living Color* (Bloomington: Xlibris Corporation, 2010), 63.

[82] Osseo-Asare, 'We Eat First With Our Eyes', 28; Frimpon, *Oversized Smock*, 283, 397.

[83] Johannes Fabian, 'Popular Culture in Africa', *Africa* 48, no. 4 (1978): 317.

[84] Frimpon, *Oversized Smock*, 216.

spices.[85] Despite their growing popularity among the working classes and the important role they performed for workers as a place to socialize and to eat traditional food, some looked askance at the frequent consumption of food at chop bars. Cheap, convenient, and necessary for survival, they nevertheless had the reputation for serving food that was not always of the highest standard (Fig. 7.3).

Fig. 7.3 Ghanaian Times Brandy Advert 7/10/1966

[85] Gracia Clark, 'From Fasting to Fast Food in Kumasi, Ghana', in *Food Consumption in Global Perspective: Essays in the Anthropology of Food in Honour of Jack Goody*, ed. Jakob Klein and Anne Murcott (Basingstoke: Palgrave Macmillan, 2014), 57.

Another reason that chop bars had an unsavoury reputation was that they reflected attitudes about workers and the current social hierarchy: "A chop bar sells mainly local meals to low-income people. Those who are well-placed socially or financially go to a restaurant."[86] Respectability being an important attribute for aspirational Ghanaians, chop bars had a problem with their image as they were distinctly working-class places to socialize. Patrons were diverse, representing locals from different communities and migrants from different countries, flaunting the transgression of barriers of ethnicity, age, and gender. This social promiscuity was subversive of the social order. Chop bars were frequently considered to be sites of moral degeneracy as they posed economic and sexual competition for wives, contesting the very definition of sexuality and the obligations of marriage. Even today it is common to hear students and bachelors referring to kenkey sellers as their wives because they depend on them for their meals.[87] Providing a taste of home by selling a variety of heavy regional foods, they also sold alcohol and were places where prostitutes solicited. If the husband felt that his proper meal was not of good quality or not served in a timely manner, he might eat more than just a proper meal at the chop bar.[88] Most coastal elites would not frequent chop bars for this reason.

Respectable people drank and ate in European style hotel bars and restaurants, or entertained at home, discreetly drinking imported alcohol and serving western foods in trusted company.[89] As cooking the proper meal became a monetized transaction, this way of cooking and eating blurred the lines of conventional social obligations between men and women, creating discomfort for traditionally minded Ghanaians. Chop bars violated the socially acceptable division of labour with the monetization of cooking for strangers.

To make matters worse for the reputation of the chop bar, some women came to generate enough income from the sale of alcohol and cooked food to expand their own businesses or to invest in colonial industries.

---

[86] Ibrahim K. Gyasi, 'Aspects of English in Ghana', *English Today* 7, no. 2 (April 1991): 28.

[87] Awo, 'Marketing and Market Queens', 150.

[88] Gracia Clark, *Onions Are My Husband: Survival and Accumulation by West African Market Women* (Chicago: University of Chicago Press, 1994), 348, 55.

[89] Akyeampong, *Drink*, 59.

This economic activity flew in the face of the very definition of sexuality and the obligations of marriage. Furthermore, economic independence generated from the business of selling cooked food and alcohol allowed women to contest their jural minor status in the courts of the day.[90] The question of who controlled women's sexuality and, by extension, their cooking labours caused many working-class men considerable uneasiness. The highlife song "Man on Top" with its assertion that women were inferior to men, verified the anxiety men were experiencing over women's growing independence.[91]

Yet another reason for the unsavoury reputation of chop bars was the traditional fear of bewitchment that many ethnic groups in Ghana maintained. For example, among the Akan certain foods were powerfully linked with the performance of witchcraft. Palm oil, an essential ingredient in southern cuisine, with its red colour and viscous consistency was considered ideal for the concealment of witchcraft substances.[92] There was also the Akan belief that one could only be bewitched by someone with whom they had an intimate physical or social relationship. Cooking for someone is an intimate act, open to the ingestion of all manner of ills for the consumer of the food. This cultural disquiet over potential food contamination gave the impression to some that the chop bar was a potentially dangerous place.

The poor reputation of chop bars aside, in an increasingly monetized economy the need of migrant workers to have a place to purchase a hot, satisfying meal forced many to abandon these fears and to turn to more western influenced ways of eating. The chop bar and restaurants became part of a new culture of eating which constituted boundaries of class identity within urban communities. They also became a symbol of new notions of modernity where eating outside the home became a potential identity marker in the political and cultural struggles between traditional village elder control and the autonomy of the wage earner in an urban setting.[93]

---

[90] Akyeampong, 66–67.

[91] E. J. Collins, 'Ghanaian Highlife', *African Arts* 10, no. 1 (1976): 67.

[92] Mensah Adinkrah, *Witchcraft, Witches, and Violence in Ghana* (Berghahn Books, 2015), 60.

[93] Sami Zubaida, 'Drink, Meals and Social Boundaries', in *Food Consumption in Global Perspective: Essays in the Anthropology of Food in Honour of Jack Goody*, ed. J. Klein and A. Murcott (Houndmills: Palgrave Macmillan, 2014), 214–16.

Foreign dishes were offered at chop bars as part of the new culture of eating in Ghana. The mining industry attracted migrant workers from all over West Africa. This diverse workforce contributed several foreign dishes to Ghanaian foodways. Nigerian workers popularized *gari*.[94] Gambian and Senegalese workers favoured *yassa*—a grilled chicken or thick fish gravy dish served with rice. There was Senegalese jollof rice, Lebanese *kebab* (small meat chunks) and *kofta kebab*—ground meat pressed around a skewer in a long patty.[95] New foods were often first incorporated by their introduction into less formal mealtimes such as the noon repast. One food that gained popularity in recent years is *acheke (attiéké)*, a fermented cassava dish from Côte d'Ivoire that is nearly black in colour and has a coarse crumb texture similar to cous cous. *Acheke* was introduced by street vendors and purchased by school children at lunch time. Foods like these then made their way into the home as women began making them for their children's midday meal.[96]

One migrant group that introduced new foods and capitalized on the adoption of Western style restaurants and supermarkets were the Lebanese. Fleeing conflict and economic disaster from the collapsing Ottoman empire in the mid-nineteenth century, by the early twentieth century the Lebanese began to settle in Côte d'Ivoire, the Gold Coast and Nigeria. Lebanese records in Ghana cite William Ibrahim Chebib and Elias el-Khoury as the earliest Lebanese to have arrived in the Gold Coast in 1884.[97] Between 1910 and 1945, the Lebanese came to play an important middleman role in west African commerce, occupying the strategic space between large European importers and exporters and west African consumers and agricultural producers. In the early days they specialized in distributing imported manufactured goods and foods and in the transport of local agricultural goods to Europe. The Lebanese in West Africa

---

[94] Fran Osseo-Asare, interview by author, 29 March 2013.

[95] Sukai Jallow, 'A Feasibility Study into the Establishment of a Gambian Restaurant in Accra' (BA Thesis, Ashesi University, Accra, 2018), 34.; Clark, 'Fasting', 58, 62.

[96] Amanda Logan, 'A History of Food Without History: Food, Trade, and Environment in West-Central Ghana in the Second Millennium AD.' (PhD diss., Ann Arbor, University of Michigan, 2013), 324–25.

[97] Emmanuel Akyeampong, 'Race, Identity and Citizenship in Black Africa: The Case of the Lebanese in Ghana', *Africa (London. 1928)* 76, no. 3 (2006): 304.

emerged as a significant entrepreneurial class, with small manufacturing, retail stores, auto trading, food production, and commercial enterprises, run predominantly by families. New waves of Lebanese came in 1960 and 1980 for economic opportunity and/or to assist their families in their various businesses.[98] By the 1980s and 90s Lebanese immigrants were positioned to respond to the need for services, including the rising demand for supermarkets, fast food eateries, and formal restaurants (Fig. 7.4).

An expanding Ghanaian middle class of technocrats and businesspeople emerged from the Structural Adjustment (SAP) policies of the 1980s under the Provisional National Defence Council (PNDC) government of Flight Lt. Jerry Rawlings. As previously discussed, the International Monetary Fund (IMF), World Bank (WB), and western capitalist aid donors set SAP's as a condition for financial assistance. Such policies involved liberalization of domestic prices (ending controls), foreign

Fig. 7.4 Butcher at Laramart Osu, Accra

[98] 'From Lebanon to Africa', accessed 29 September 2020, https://www.aljazeera.com/program/al-jazeera-world/2015/10/28/from-lebanon-to-africa/.

exchange prices (devaluation of currency), and trade. SAPs were meant to incentivise exports and to curtail demand for imports.[99] These neoliberal policies amplified income polarization and intensified pressure on poor and lower middle-class women to work longer hours to generate income. Many poorer families were pushed out of central neighbourhoods to distant suburbs where travel times were multiplied. Many suffered from the severely depressed economy which resulted from these policies, but simultaneously SAPs promoted a period of sustained economic growth. This growth had the effect of enlarging the middle class of Ghana.

A larger middle class consisted of educated people who were employed by both public and private enterprises and businesspeople, such as petty retailers and importers.[100] In recent years, this emerging middle class has been characterized by families where both parents worked in corporate and/or trading industries. Working mothers gained breathing room by shifting breakfast and lunch into the category of individual purchases. Lunch money could be left for children at home to purchase food from commercial vendors.[101] By the 1990s, many chose to assign proper meals to Sundays when not as many people had to work. Weekday evening meals have been supplanted by fast foods such as kenkey or fried rice which has largely replaced more nutritious, less salty, less greasy jollof rice. Convenience has become more important as parents needed to be located where the jobs were and extended families no longer lived near each other.[102] Despite this reorganization of eating habits in urban areas, gender remained a primary structural element in the meanings assigned to food. Women were still responsible for providing sustenance for their families. The marketing of convenience foods such as powdered fufu remained strongly gendered with images of women emblazoned on local convenience products.[103]

Immigrant workers continue to drive the expansion of European style bars, large hotels, and restaurants. Émigrés from China working on

[99] Jon Kraus, 'The Struggle over Structural Adjustment in Ghana', *Africa Today* 38, no. 4 (1991): 19.

[100] Ian E. A. Yeboah, 'Structural Adjustment and Emerging Urban Form in Accra, Ghana', *Africa Today* 47, no. 2 (2000): 79–80.

[101] Clark, 'Fasting', 55.

[102] Osseo-Asare, interview by author.

[103] Clark, 'Fasting', 58.

development projects ushered in an increase in Chinese food eateries in the Osu district of Accra and in state run hotels such as Labadi Beach or the Golden Tulip. The same is true for the increasing availability of Indian, Ethiopian, and Mediterranean food, demand for which is driven by workers on western salaries.[104] Lebanese entrepreneurs, many of whom were born and raised in Ghana, capitalized upon this growing demand with investments in fast food restaurants, butcher shops, and restaurants that sold Lebanese and other foreign foods. Many Lebanese families managed western style supermarkets and owned fast food franchises.[105]

Twentieth Century Adoptions and Adaptations

Colonial capitalist production in the late nineteenth and early twentieth centuries encouraged the importation of food. Workers flocked to urban centres such as Takoradi, Accra, or Kumasi, and imported tinned foods offered a solution to meet growing demand for food. One official remarked in 1921 that "the tendency of the literate African to prefer tinned articles of food to his native foodstuffs is universal".[106] Another comment by an English colonial official in 1928 was similarly revealing: "The people of Ashanti are more likely to be ruined physically, morally, and financially by tinned foods than by strong drink."[107] These commentaries were strong testaments to the popularity of tinned foods and their affordability for wage earners. Recipe books produced at the time contained advertisements from grocers like Lazenby's who supplied the Gold Coast with "potted" meats, fish, and salmon, as well as wheat flour, oatmeal, and rice.[108] Imported foods helped workers to fill a gap in towns and cities that were experiencing difficulty keeping up with rising demands for food.

---

[104] Osseo-Asare, interview by author.

[105] S. O. Marfo, 'Immigrant Business in Ghana. A Study of the Lebanese in Accra' (MPhil Thesis, University of Ghana, 2012).

[106] Gold Coast Government, 'Report on the Census', 1921, ADM 5/2/5, PRAAD Accra.

[107] Kumasihene, 'By-Laws Concerning Consumption of Liquor at Funeral Customs', 1928, ARG 26/8/12, PRAAD Accra.

[108] *The Gold Coast Cookery Book* (Accra: British Red Cross, 1933).

Some of the most popular tinned items were meats such as sardines, corned beef, and bologna sausage. Quick and convenient, market women in colonial times and after independence served these meats with bread to wage earners in search of a quick meal, and bread and sardines were widely consumed by travellers unfamiliar with the local foods.[109] By far the most often consumed tinned meat was the humble sardine. As already stated, sardines could be readily incorporated into local soups and stews. In 1937, imported sardines and stock fish were listed as "articles for personal consumption" by cocoa farmers in the central forested zone.[110] There was little in the local repertoire that one could carry to work for breakfast or lunch other than the raw ingredients themselves.[111] Tinned sardines and other meats gained a foot hold in the local foodways by providing an affordable source of protein that was also more transportable.

Tinned or potted foods like sardines were in such high demand that they were smuggled into the colonial Gold Coast to avoid the duty. The procedure in 1943 was to send smuggled items to police in Sunyani, district headquarters for the colonial government in today's Brong Ahafo Region. Sunyani was an important hub for the distribution of cocoa, kola nuts, and staple foods such as maize and yams. Smuggled items were to be sold at a public auction, the proceeds of which were meant for the maintenance of the police force there. But in 1944, the policy had changed to have them forwarded to the Comptroller of Customs in Accra for disposal. This caused complaints from Sunyani, whose staff keenly anticipated the sardines: "Last week, 3 or 4 cases of sardines were sent to Kumasi by the Police—a luxury almost unknown in this District and one which would have been bought at once."[112] By the 1950s, government

---

[109] Thompson, 'Inspector General of Police to Commissioner of Police Ashanti', 1928; Goody and Goody, 'Food and Identities', 10.

[110] Great Britain and Commission on Marketing of West African Cocoa, *Report of the Commission on the Marketing of West African Cocoa.* (London: H. M. Stationery Off., 1938).

[111] Goody and Goody, 'Food and Identities', 10.

[112] District Commissioner, 'Sunyani District Commissioner's Letter to Kumasi Chief Commissioner on Confiscated Goods', 19 December 1944, ARG 1/1/221, PRAAD Kumasi.

surveys added sardines to the most frequently purchased items, especially in the cities.[113]

By the late 1960s tomatoes also became an essential cooking ingredient in sauces and soups.[114] By the mid-twentieth century, tomatoes together with tinned fish and onions made their way into the main meal of the day. Ghanaian cookbook author Dina Ayensu stated in her 1972 cookbook that "Onions, tomatoes, salt, and pepper, the minimal fundamental ingredients of West African dishes, must be carefully used in correct proportions."[115] Tinned tomatoes became a way to subvert the seasonal availability of fresh tomatoes.[116] So ubiquitous had the use of tomato become in local cooking that with very few exceptions by the 1960s they were used in most soups as a thickener and a way to add the desired red colour customarily derived from palm oil.

The tomato had become an accepted and preferred convenience ingredient to augment labour-intensive palm oil recipes in southern Ghana. There was considerable demand for fresh tomatoes from the commercial sector as well. Large quantities of tomatoes are sold to restaurants and commercial food sellers making *kenkey*. The fresh tomatoes which are a key ingredient in kenkey cannot be substituted with imported paste. For that reason, these groups of food sellers need a constant supply of fresh tomatoes.[117] Fresh tomatoes are grown domestically—mostly in the northern Ghana—and traded all over the country. Burkina Faso is an important source for both fresh and tinned tomato paste.[118] It is estimated that 90 percent of the tomato paste consumed today in Ghana

[113] Office of the Government Statistician, *Accra Survey of Household Budgets* (Accra: Government Printing Department, 1953), 2.

[114] Dede, *Ghanaian Favourite Dishes*, 2–8; Ayensu, *The Art of West African Cooking*, xiv; Goody and N. Goody, 'Food and Identities' 10; Clark, *Onions Are My Husband*, 352.

[115] Ayensu, *The Art of West African Cooking*, xiv.

[116] Logan, 'History', 80.

[117] Awo, 'Marketing and Market Queens', 150.

[118] Ibid., 126; Joseph Amikuzuno, 'Border Effects on Spatial Price Transmission between Fresh Tomato Markets in Ghana and Burkina-Faso: Any Case for Promoting Trans-Border Trade in West Africa?', in *85th Annual Conference* (Agricultural Economics Society, Warwick University, Coventry, UK, 2011).

comes from the European Union.[119] As tomatoes became indispensable to cooking and eating in Ghana, their production and trade became big business and remain a political issue that is intently scrutinised by scholars and policy makers.[120]

Unlike sardines and tomatoes, sugar was not an ingredient that could be incorporated into the main meal of the day. Instead, sugar found its niche at morning and midday mealtimes when standards for what was considered proper food were considerably relaxed. Missionary and government schools helped to expose many to the taste of sugar. Contributors to *The Gold Coast Cookery Book*, published in 1933, included two local nuns, Sister Angele and Sister Antonia of Our Lady of Apostles, a Catholic mission school for girls in Cape Coast, founded in 1890. Fully five out of the 12 sections featured sugar as the main ingredient with sections entitled "Hot sweets", "Cold sweets", and "Confectionary".[121] The desire for sugar in numerous cake and bread recipes, initially fed by the combined influence of Islam and Christianity, proliferated with colonization and westernized ideas of what constituted a proper breakfast, which often included wheat and sugar.

In the 1940s, sugar consumption took on an aspirational and performative character for many Ghanaians. Both the 1942 and 1944 Fun Fairs for War Charities in Kumasi's Jackson Park proved the degree to which sugary foods had become associated with special occasions. Under the joint patronage of the Chief Commissioner of Ashanti and the Asantehene, Sir Osei Agyeman Prempeh II, treats and games were organised to raise money for charity: "Excellent tea with homemade cakes will be provided at moderate prices. Minerals [sodas] and other drinks will be sold at the

119 Martha A. Awo, *Marketing and Market Queens: A Study of Tomato Farmers in the Upper East Region of Ghana* (Berlin: LIT Verlag Münster, 2012), 16.

120 Fergus Lyon, 'Trust, Networks and Norms: The Creation of Social Capital in Agricultural Economies in Ghana', *World Development* 28, no. 4 (1 April 2000): 663–81; Justice Akpene Tambo and Theresa Gbemu, 'Resource-Use Efficiency in Tomato Production in the Dangme West District, Ghana', in *International Research on Food Security, Natural Resource Management and Rural Development* (Tropentag, Zurich, 2010), 5; Samuel Donkoh, Mark Tachega, and Nelson Amowine, 'Estimating Technical Efficiency of Tomato Production in Northern Ghana', *American Journal of Experimental Agriculture* 3 (December 2012): 56–75; 'Interview with Dr. Kofi Amoah', *State of Affairs* (Ghana: GHOne, 2 December 2016). This is but a small sampling of the scholarly articles and debates produced around the cultivation and trade of tomatoes.

121 *The Gold Coast Cookery Book,* 115, 139, 265.

bar."[122] The appearance of the Asantehene and the attendance of all classes of African, European, and Lebanese communities, combined with the serving of sugary confections and drinks, elevated this event to a performative occasion of regional Asante pride and wealth.

Sugar proved to be accessible to all class levels, even as it was being used to signal wealth and prosperity by elites. By the late 1940s, sugar was available as a street food consumed by children, albeit in its raw form. Busia's 1947 social survey examined the family and occupational structure in Sekondi-Takoradi and how this affected the nutritional wellbeing of school children. After the administration of surveys to the children, Busia learned that many children had to travel quite far to attend school and were unable to go home for their midday meal. The little money that most children had to purchase food (usually two pence) "…is generally put into starchy foods, mostly garri…and often the money goes into sugar-cane and sweets".[123] School children were exposed to daily doses of imported sugar, creating a dependency of taste that would grow over time as they did.

In the 1950s, the aspirational aspect of sugar grew with the marketing of imported beverages, such as Horlick's and Ovaltine drinks advertised in urban centres. Marketing imagery invoked impressions of professionalism and pride for lorry drivers and delivery men who would benefit from the convenience, health, and strength that a morning serving of Ovaltine could provide. Sugar was further popularised in the 1950s by market women who sold baked, sweet tea bread to lorry drivers and children.[124] By the 1960s, baking with sugar was widely embraced by every class of the local population. Newspaper articles touted European recipes for German Christmas cookies, cakes, and tarts.[125] Sugar fully embodied the drive and the ambitious mood of an independent Ghana, and the Tate and Lyle refinery of Britain was there to assist. In October of 1966, it was announced that the following February, the Accra International Trade Fair would be supplied with "enough sugar cubes to satisfy the requirements of fairgoers". Individually wrapped sugar cubes bearing the

[122] 'Fun Fair in Aid of War Charities', 19 February 1942, ARG 1/1/215, PRAAD, Kumasi.

[123] Busia, 'Sekondi Social Survey', 51.

[124] Goody and Goody, 'Food and Identities', 7.

[125] 'Hints for Christmas Cooking', *The Ghanaian Times*, 24 December 1968, PRAAD Accra; 'When You Get Visitors…', *The Ghanaian Times*, 7 October 1966, PRAAD Accra.

fair logo were provided free of charge by Tate and Lyle. Not to be outdone, Kingsway Stores (a subsidiary of the Dutch multinational Unilever) was to serve the accompanying tea, coffee, minerals (sodas), lagers, wine, and spirits to attendees.[126] Ads with chubby-cheeked children and happy, smiling adults gleefully consuming imported, sugary Ovaltine biscuits signalled that this product was meant to produce healthy, prosperous families (Figs. 7.5 and 7.6).

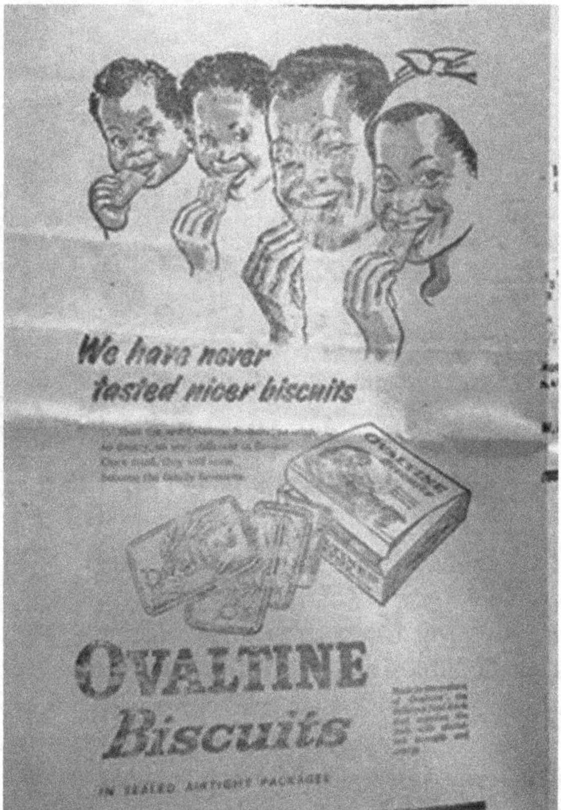

Fig. 7.5 Ghana Times, 8 October 1966

[126] 'Free Supply of Sugar at Trade Fair', *The Ghanaian Times*, 10 October 1966, PRAAD, Accra.

Fig. 7.6 The Daily Echo, 21 May 1954

By the late 1960s, sugar imports reached an all-time high and caused a bit of a crisis, as the subsidising of this import was depleting the national budget. In addition to this, the government was having trouble battling smugglers. Attempts were made to substitute imports with locally grown sugar processed at the state owned Komenda Sugar Factory.[127]

[127] 'Our Sugar Problem Can Be Solved If…', *The Ghanaian Times*, 18 October 1968, PRAAD Accra; 'Bank to Grant Loans to Sugar-Cane Farmers', *The Ghanaian Times*, 17 December 1968, PRAAD Accra; 'Police Seize 600 Cartons of Sugar', *The Ghanaian Times*, 17 October 1968, PRAAD Accra.

A change in attitudes toward sugar was detectable in cookbooks of the period. Some cookbooks and newspapers offered recipes for Western baked foods and sweets.[128] Others chose not to add sugary items to their recipes, stating that desserts were not a customary part of West African foodways and that "Such desserts as puddings, ice cream, cakes, pastries, and fritters that we do have are traceable to the influence of the West on our culture".[129] Despite the disapproving tone that began to develop in the late 1960s, the data on food trade in Ghana indicated that sugar importation was on an upward trend and continued to grow through to the 2000s.

***Bodobodo*—Twi, *Bodobodo*—Gã, *Apono*—Fante**

Imported wheat flower
Sugared Palm Wine

- Mix the flour and sugared palm wine into a dough and allow to stand for six hours.
- Knead and leave another hour.
- Shape into a loaf and leave for another twenty minutes.
- While the dough is resting, heat a hemispherical oven with a faggot of wood. Once the greatest heat has been reached, rake out the ash.
- Place the risen loaf in the hot oven and allow to "soak" until the bread has risen and crusted.

Adapted from Field, 'Gold Coast Food,' 14.

Like sugar, western style bread occupied the niche in the meal scheme that was not already occupied by the proper staple served at the main meal of the day. European-style bread was adopted from its introduction by the Portuguese on the coast in the fifteenth century and grew in popularity in the nineteenth century.[130] It was adapted with the addition of palm wine yeast in the mid-nineteenth century.[131] In the early twentieth century it was still being made with fermented palm wine.[132] The addition of palm wine

[128] 'Hints for Christmas Cooking'; Dede, *Ghanaian Favourite Dishes*; 'When You Get Visitors…'

[129] Ayensu, *The Art of West African Cooking*, xix.

[130] J. D. La Fleur, *Fusion Foodways of Africa's Gold Coast in the Atlantic Era* (Leiden: Brill, 2012), 117.

[131] Richard Francis Burton, *Wanderings in West Africa from Liverpool to Fernando Po* (London: Tinsley brothers, 1863), 144.

[132] Decima Moore Guggisberg, *We Two in West Africa* (London: W. Heinemann, 1909), 225.

yeast made the local approximation of bread distinctively sweet and dense. A large loaf would serve as a mid-day meal eaten by one person without any other food.[133] By 1917, especially if you were a scholar at a European school or a clerk, the habit of eating bread and drinking tea for breakfast had become pervasive. This was in part because bread was cheaper to produce than kenkey in urban areas.[134] Efforts to curtail the consumption of wheat bread after the First World War by colonial administrators were unsuccessful as it had become part of the local repertoire: "many Ghanaians embraced 'foreign' foods like white bread and tinned meats as part of the national cuisine, frustrating colonial efforts to reform foodways."[135] By 1950, a survey of Accra found that bread had become a daily ritual and that it was produced mainly outside the home, much like kenkey was outsourced: "that bread was being consumed at least once a day…, and that the main sources of bread supply were either local bakeries or itinerant market women: in no recorded case home baking served only household consumption."[136] Bread was popular in other urban areas as well. Children in Sekondi-Takoradi worked in the growing informal economy by selling imported biscuits, locally produced breads, and snack foods to supplement the family food budget.[137]

Colonial administrators resorted to conservation arguments in combatting the daily consumption of bread in urban areas, predicting the extinction of the palm tree:

> To make the palm-wine, the tree is commonly cut down in a most wasteful system, a tree which may have taken 50 years or more to grow being destroyed for the sake of a few days' palm-wine … The alcoholic content of the sap, after 7 days' fermentation, varies from 2.3 per cent to 5.1 per cent … It is commonly used as yeast in local breadmaking.[138]

[133] Field, 'Gold Coast Food', 14.

[134] 'Agricultural Mutual Improvement Association', *Gold Coast Nation*, 7 April 1917.

[135] Robins, 'Food Comes First', 169.

[136] Gerardo Serra's discussion of PRAAD RG 4/1/124 Survey of bread consumption in Accra in, 'Towards a Political Economy of Statistics: A Study of Household Budget Surveys in the Gold Coast, 1945–1957', in *Session on Measurement, Planning and the State in Sub-Saharan Africa: Historical Perspective'* (Vancouver: Simon Fraser University, 2013), 14.

[137] Busia, 'Sekondi Social Survey'.

[138] Frederick Robert Irvine, *Woody Plants of Ghana: With Special Reference to Their Uses* (Oxford: Oxford University Press, 1961), 777–79.

The popularity of bread however, continued to spread beyond urban centres. Wheat bread, along with tea, tinned sardines, tomatoes, and evaporated milk were commonplace even in rural areas of Ghana.[139] Field work in the Western and Asante region in the late 1980s found that as many as 24 percent of respondents had a combination of bread with porridge or a hot drink instead of leftover kenkey, and fish or rice and stew for breakfast.[140] Convenient, making breakfast easy and fast to prepare, bread came to displace leftover traditional foods as the first meal of the day in Ghana (Fig. 7.7).

Today the Ghanaian government is investing in research into the substitution of wheat flour with native starches in the interests of boosting consumption of local foods. At the Food Research Institute (FRI) in Accra, I was treated to a sample of bread made with yam flour. One of the mandates of the FRI is to work with entrepreneurs to find different ways

Fig. 7.7 Bread Oven Kaneshie, Accra

[139] Jack Goody, *Cooking, Cuisine and Class: A Study in Comparative Sociology* (Cambridge: Cambridge University Press, 1982), 176.

[140] Osseo-Asare, 'We Eat First', 53; Clark, *Onions*, 356.

to develop, package, and market local foods.[141] Irrespective of these efforts, the popularity of wheat bread was a breakfast fixture for locals and for travellers. In marketplaces, and especially in transit loci like bus stations, in the middle of rush hour traffic, or on the road from Accra to Kumasi, young people can be seen selling bread and biscuits to travellers.

One of the most significant changes in the mid-twentieth century was the surging consumption of imported rice from places such as the United States and Thailand. Since the 1950s, rice has become a standard food of middle-class government workers and teachers, who consume imported American rice.[142] Over the last three decades, Asian rice (*Oryza sativa*) has almost completely supplanted African red rice (*Oryzaa glaberrima*) because it is quick to prepare, and it has become the food of urban dwellers in Ghana (Fig. 7.8).[143]

Another popular dish for city dwellers in Ghana is jollof rice, a heavy single pot dish that combines rice with meats and seasonings and derives its name from the Wolof people of the Jolof Kingdom of present-day Senegal.[144] In the mid-fourteenth century, the Wolof broke away from Mali to form the Jolof Empire which gave political, cultural, and linguistic unity to the Wolof and enabled them to attain the predominant position they presently hold in Senegal today.[145] The Senegalese version of jollof rice is called *thieboudienne* or *cheebuyapp.*[146]

Senegal is part of the long-established region of rice cultivation and consumption on the upper Guinea coast, an historical term in use from the sixteenth to the nineteenth century to describe a coastal area which covered all or parts of present-day Gambia, Senegal, Guinea-Bissau, Guinea, Sierra Leone, Liberia and Cape Verde.[147] Rice has been cultivated in the hinterland of the upper Guinea coast, between the upper

---

[141] Miller, 'Food and Nationalism", 95.

[142] Goody and Goody, Food and Identities, 5.

[143] Logan, History, 11.

[144] Osseo-Asare, *Food Culture*, 33.

[145] Eunice A. Charles, *Precolonial Senegal: The Jolof Kingdom, 1800–1890* (Boston: Boston University African Studies Center, 1977), 1.

[146] McCann, *Stirring the Pot*, 133.

[147] Jacqueline Knörr and Christoph Kohl, *The Upper Guinea Coast in Global Perspective* (Oxford: Berghahn Books, 2016), 1.

Fig. 7.8 Texas and Thai rice, Palace Supermarket, Accra

Senegal and upper Niger Rivers, since before the period of Islamic expansion into the region in the tenth century.[148] Current evidence establishes the cultivation of native African rice—*Oryza glaberrima*—along the middle Niger in Mali some two thousand years ago.[149]

Jollof rice is actually a hybrid version of *thieboudienne*—or *benachin* as it is called in the Gambia. Versions of jollof rice have diffused from Senegambia all the way to Nigeria. Regardless of what the traditional local starchy staple was, jollof achieved popularity with urban cultures throughout West Africa. African food historian James McCann posited an hypothesis about the presence and popularity of jollof rice in the urban settings of West Africa. The Djula people are a Mande trading group inhabiting several West African countries, including present day

[148] Judith A. Carney, *Black Rice: The African Origins of Rice Cultivation in the Americas*, New Ed edition (Cambridge, Mass.: Harvard University Press, 2002), 33.

[149] D. W. Phillipson, *African Archaeology* (Cambridge: Cambridge University Press, 2005), 203.

Mali, Côte d'Ivoire, Ghana, and Burkina Faso. Since the fourteenth century the Djula established trading communities in urban centres all over West Africa. McCann theorized that the Djula manifested their presence in local cooking cultures, appropriating new ingredients from local women, resulting in hybridized versions of jollof all over West Africa.[150]

Jollof Rice

1 lb. rice
1 lb. meat
Onions
1 ½ lb. tomatoes
½ lb. lard
Pepper and salt

- Cut the meat into pieces, wash and put to boil.
- Prepare the onions and tomatoes and grind the pepper.
- Remove meat when tender, drain, and fry in hot lard until brown.
- Fry onions, tomatoes, and pepper until cooked. Add fried meat and the water in which it was boiled. Add salt to taste and boil for some time.
- Wash the rice and add to the stew. Add a bit of water when necessary to keep the rice moist. Serve hot or cold.

Adapted from *Ghanaian Favourite Dishes*

Debates over which West African country has the best jollof rice have grown fierce, fuelled by the variety of ways it is prepared, dependent upon the availability of local ingredients: "This variation in recipes is perhaps at the root cause of this debate, which is a popular topic not only in social media, but also in street conversations."[151] The variations are also determined by whether or not the rice is parboiled (partially boiled in the husk), or if it is cooked in one pot, over firewood or charcoal. Even Mark Zuckerberg, CEO of Facebook, on a visit to Nigeria in 2016 and the Nigerian information minister, Lai Mohammed, in 2017, have been

[150] McCann, *Stirring the Pot*, 135.

[151] Anisa Subedar and Iqbal Ahmed, 'West Africa Steams over Jollof Rice War', BBC News, 26 August 2017, https://www.bbc.co.uk/news/blogs-trending-41053424; Ofeibea Quist-Arcton, 'Jollof Rice: West Africans Dish It Up With A Hefty Serving Of Smack Talk', NPR.org, accessed 10 September 2019, https://www.npr.org/sections/thesalt/2016/08/30/491380208/jollof-rice-west-africans-dish-it-up-with-a-hefty-serving-of-smack-talk?t=1568145970707.

caught up in the controversy, the latter receiving calls for resignation after he carelessly stated that Senegal had the best jollof rice. Opinions are so strong that Zuckerberg's handlers advised him ahead of time, and he wisely stated that he was told not to compare Nigerian jollof to that of other countries.[152]

As stated, imported rice was tasty, satisfying, and convenient in that it cooked rapidly and required comparatively (to maize, millet, and yam) no processing. Recipes using rice incorporated local and imported ingredients, such as black-eyed peas, raisins, groundnuts/peanuts, frozen okra, garden egg (eggplant), tinned lobster, mushrooms, and boiled eggs. In addition to tomatoes and rice, instant stock cubes began to be adopted as a convenient ready seasoning as early as 1933. *The Gold Coast Cookery Book* specified the use of oxtail, vegetable, or mulligatawny flavoured packets in a soup to which was added items like minced meat, fried onion, and was thickened with wheat flour.[153] Not all cookbooks recommended the use of instant soup cubes and wheat flour. In *Ghanaian Favourite Dishes*, onion and/or tomato were used exclusively as seasonings in every single soup recipe, with no mention of seasoning cubes. Recommended thickeners included traditional dried okra or dried baobab leaves.[154] However, in *The Art of West African Cooking* a flavour enhancer named *Ac'cent* was listed as a seasoning ingredient in soups.[155] Odourless and composed of maize glutamate, which provided a savoury umami flavour to soups and stews, *Ac'cent* began to be marketed in 1947.[156] These seasonings quickly became requisite to local cooking, though not without some resistance.

In more recent years, corporations like Maggi—a subsidiary of Nestlé—had an increasing market share due to inexpensive, convenient products such as seasoning cubes. Seasonings, or bouillon cubes, were made to be especially convenient and adaptable because they could be

[152] 'Nigeria Jollof Delicious—Mark Zuckerberg', accessed 28 April 2020, https://www.ghanaweb.com/GhanaHomePage/entertainment/Nigeria-jollof-delicious-Mark-Zuckerberg-466952.

[153] *The Gold Coast Cookery Book*, 49.

[154] Dede, *Ghanaian Favourite Dishes*, 2–8.

[155] Ayensu, *Art of West African Cooking*, 4.

[156] B&G Foods, Inc., 'Ac'cent Flavor Enhancer', Accent Flavor, accessed 2 July 2019, https://www.accentflavor.com/about.

purchased in individually wrapped cubes, which are affordable for most people. In northern Ghana in the late 1960s for example, increased reliance upon the market due to a decline in cultivation, foraging, and hunting has caused a more noticeable increase of seasoning cubes to impart a protein flavour to soups instead of the more traditional *dawadawa* seeds.[157] Asantewaa Tweedie, author of *Melting Pot: Ghanaian Cooking with a Twist*, commented on the use of these seasoning cubes:

> A good stock is the basis to any good soup or stew. However, nowadays people view the process as too time consuming, and frequently use artificial or convenient seasonings, commonly sold in cubes or powder form… Most, if not all of these artificial seasoning, put MSG (mono-sodium glutamate) and other artificial flavorings in them. Many people have adverse reactions to these artificial ingredients such as swollen tongues, sore gums, nausea, stomach ache, indigestion and so on.[158]

Tweedie's commentary reveals the reason for disapproving attitudes of some Ghanaians toward convenient seasoning cubes and reflects the desire of some to maintain the integrity of regional cuisines. In Tweedie's cookbook, foreign food was perceived as the unhealthy other, while local food is thought to be life sustaining and healthful. Regardless of the attitudes of some toward ready seasonings, Maggi seasoning cubes continue to be widely available for purchase on the street and in stores and markets. Their popularity rested upon the ease with which they could be incorporated into the main meal of the day.

Another food that could easily be incorporated into the main meal of the day was cassava in the form of *gari*. Indeed, *gari* was a food much coveted by city dwellers at the beginning of the twentieth century, as it was cheaper than most other foods, easier to prepare, and more convenient to store, taking up less space in the pantry than other staples.[159] The interwar period was a time when both cassava and rice became especially

---

[157] Jessica R. Ham, 'Cooking to Be Modern but Eating to Be Healthy: The Role of Dawa-Dawa in Contemporary Ghanaian Foodways', *Food, Culture & Society* 20, no. 2 (2017): 246, https://doi.org/10.1080/15528014.2017.1305827.

[158] Tweedie, *Melting Pot*, 3.

[159] Ohadike, 'The Influenza Pandemic', 390.

popular. Considered to be a "prepared food", *gari* was introduced to Accra in the interwar years by migrants. Fish women—women who smoked fish and then sold it long distance and locally—found the sale of *gari* to be a lucrative side-line, along with other imported provisions such as sugar, sardines, wine, and beer. By the early 1970s, fish, as a trade item in Accra, was declining relative to the trade in staples such as cassava (processed into *gari*) and maize.[160] Irrespective of its popularity with other ethnic groups such as the Gã and Ewe, with those from the Akan ethnic group, cassava was an ambivalent blessing due to its negative connotations as food for slaves. The following is a statement about cassava from a man who resided in Kwawu, on the far western coast, from the opening decade of the twentieth century to independence in 1957: "These days, food habits have changed considerably among Akan people. Owing to the high prices of plantains, much cassava is planted and consumed in Kwawu."[161] For those from the Akan ethnic group cassava was still an unacceptable comestible until the late twentieth century.

A shift to maize occurred in the central forested zone in the late twentieth century, where it was consumed as a breakfast, lunch, or snack food. In the 1980s, coastally produced maize-based kenkey was sold in Kumasi in the central forested zone. Traders travelled to Mankessim, just inland from the coast, to purchase bulk supplies of kenkey and dried fish from the Fante coast for sale for breakfast or lunch: "A very wide variety of steamed, fried, baked or boiled items of diverse ethnic origins are relished for lunch and snacks by Asantes very conservative about their evening meal."[162] These such items included tea and bread (European), *gari* and beans (Ewe), *kenkey* and fish (Gã or Fante), and boiled yam or plantain with stew. But these other starches still did not displace the main meal of the day. In fact, the activity of market traders revolved around the preparation of the main meal of the day. Often women packed up at around three in the afternoon to allow enough time to purchase ingredients for and to cook their proper meals.[163]

---

[160] Robertson, *Sharing the Same Bowl?*, 95, 111, 172.

[161] Kyei, *Our Days Dwindle*, 126.

[162] Clark, *Onions*, 356.

[163] Ibid., 349–51; Clark, 'Money', 330.

Red-Red

2 c Black Eyed Peas, uncooked
½ c Red Palm Oil
2c Chopped Onions
½ tsp Dried Ground Red Pepper
2 tsp Grated Fresh Ginger
3 Cloves Crushed Garlic
2 tsp Dried Shrimp Powder
2 c Pureed Tomatoes
2 Tbs Tomato Paste
½-1 lb Smoked Fish

- Clean peas, soak overnight, and cook until tender, about 1 hour.
- Sautee onion in palm oil.
- Add pepper, ginger, garlic, shrimp powder, & fry a few minutes more.
- Add tomatoes and tomato paste and fry together for a few minutes.
- Break fish into pieces and add to stew. Add drained peas & simmer.
- Add water if needed & check seasoning, adjusting to taste.
- Sprinkle with gari and serve with hot fried ripe plantain.

Adapted from Osseo-Asare & Baeta, 'The Ghana Cookbook', 124.

Unlike the Asante, the Ewe embraced the consumption of maize in the twentieth century as part of their main meal. Indeed, maize is now often eaten three times per day, the flavour of which is varied by the packaging in corn husks, plantain leaves, or cassava leaves. Ewe maize can be prepared fermented or unfermented and then either roasted and ground into a light or stiff porridge. The most common dish associated with the south-eastern Ewe is *akple* and okra soup. A stiff porridge made of unfermented maize and cassava dough, *akple* is cooked with continuous stirring with heat. Once sufficiently thickened, it is shaped into a circular or oval shape. Ewe *akple* is often combined with a third cassava dough to two thirds maize dough which gives it a characteristic softness.[164]

One constant in Ewe foodways discussed in both Chaps. 1 and 2, is the consumption of beans, which can be traced back to their seventeenth-century migration to the south-eastern coast. The very Ewe identity is

[164] Florence Efua Dovlo, 'Nutrition and Food Utilization', in *Handbook of Eweland: The Ewes of Southeastern Ghana v. 1*, ed. Francis Agbodeka (Accra: Woeli Publishing Services, 1997), 195; Fran Osseo-Asare and Barbara Baëta, *The Ghana Cookbook* (New York: Hippocrene Books, 2015), 189.

tied in with the eating of beans. In the early twentieth century, they were disparaged for this aspect of their cuisine: “Hitherto the greatest eaters of beans have been the Ewes whose nickname jeers at this practice.”[165] Beans remained a prominent feature of the foodways of the Ewe in the late twentieth century.[166] Cowpeas (black eyed peas) tolerated sandy soil and low rainfall and were ideal for cultivation in the microclimate they inhabited on the coast. Bambara, groundnut, and melon seeds were also foundational to the cuisine.[167] *Bobo*, a breakfast dish for workers or school children, was customarily served as an informal meal at home on the weekends. White or brown cowpeas were cooked until soft, then seasoned with salt and palm oil and served with *gari*. This bean stew was served with fried plantain pancake or *tatale*. In the early nineteenth century, *tatale* was also known as *tartari* or *tatali*. This was prepared with pounded ripe plantain combined with ground rice, onions, and salt and pepper. It was made into small cakes and fried in palm oil. In the late twentieth century, it was prepared with a dough of pounded ripe plantain and maize meal (Fig. 7.9).[168]

A variation of this bean stew is known as *red-red*. This popular bean stew dish represents the coming together of palm oil, smoked and dried fish, beans and plantains in what is now a classic Ghanaian dish.[169] This bean stew is served with fried ripe plantains and today, it is prepared with pureed tomatoes and tomato paste. Named *aboboi* or *asedua* in the late sixties and 1970s, the current recipe name *red-red* makes use of reduplication—the red of the palm oil and the red of the tomatoes—to highlight its red colour.[170] *Red-red* remains a favourite among boarding-school children and is almost always eaten everywhere with fried plantains.

Children living away from home in boarding schools found shortcuts to producing quick fermented starches such as *abolo*. Another quick but

---

[165] Field, ‘Gold Coast Food’, 10.

[166] Francis Agbodeka, ed., *Handbook of Eweland: The Ewes of Southeastern Ghana*, vol. 1 (Accra: Woeli Publishing Services, 1997), 10.

[167] Ibid., 198; Field, ‘Gold Coast Food’, 21.

[168] Agbodeka, *Handbook of Eweland*, 1:198; Field, ‘Gold Coast Food’, 16.

[169] Steven J. Salm and Toyin Falola, *Culture and Customs of Ghana* (Westport, Conn: Greenwood Press, 2002), 109.

[170] Osseo-Asare and Baëta, *Ghana Cookbook*, 124; Dede, *Ghanaian Favourite Dishes*, 42; Ayensu, *The Art of West African Cooking*, 121.

Fig. 7.9 Red-red with fried ripe plantain

fortifying porridge credited to Ewe boarding students is made of *gari* soaked in water with added sugar and milk (if milk was to hand), served with roasted groundnuts or dry coconut. The *gari* would swell to three times its size once reconstituted, and the peanuts and milk provided protein.[171] These foods are popular almost the entire length of Ghana today, and are available at most roadside rest stops, chop bars, and in restaurants all over Accra. Transcending ethnic and national boundaries, *gari's* utility was the reason for its ubiquity in Ghana and neighbouring Nigeria.

Conflicting Attitudes

It is apparent that Ghanaians are proud of their food's wholesomeness, regularly claiming that it is organic and not genetically modified. In the 12th–26th May 2014 edition of the *Your Neighborhood* newspaper, the organisation Food Sovereignty Ghana (FSG) and the Coalition of Farmers Against Genetically Modified Organisms (COFAM) were on a

[171] Agbodeka, *Handbook of Eweland*, 1:197.

mission to lead a "movement (that) is also aimed at protecting our food supply, local farms, and environments, promoting organic solutions, and exposing cronyism between big business and the government."[172] COFAM was protesting against pending legislation that would allow the Monsanto Company, a publicly-traded American multinational biotechnology corporation and a leading producer of genetically-engineered seed, to bring genetically modified products into Ghana.

Foreign foods seem to be perceived as problematic for the economy, contributing to insecurity due to ever fluctuating pricing, and representative of neo-colonial domination by others. Owutse Noah, a student in the Family and Consumer Sciences Department at the University of Ghana at Legon, preferred to eat food he brought from home prepared by his mother, or rice he prepared himself in his room instead of purchasing prepared food or eating at a restaurant.[173] He and his classmate, Agbeve Worlali, both agreed that pizza and other such foreign foods were generally considered snack foods and that Ghanaian food was "good for the body" by comparison. Additionally, both students stated that these foods may be available, but "purchasing power" was an obstacle to frequent consumption.[174] Many of the people interviewed, such as Charity Dzigbede, receptionist at the Baptist Guest House in the Cantonments neighbourhood of Accra, exhibited a suspicious attitude toward foreign food. Ms Dzigbede expressed the belief that local food was healthiest and would not contribute to disease. Adeline Doe, an Ewe cook at the Baptist Guest House, was also of the opinion that the food the Ewe consumed in the Volta region was best for digestion.

Whatever the opinion of the health benefits of local foods, the British influence was ever present in items such as tinned baked beans, tea, sausages, biscuits, and the ubiquitous meat pie (a Cornish-style pastry commonly sold by street hawkers). It was becoming increasingly common to be served rice and beans garnished with spaghetti (macaroni) in any establishment near areas where businesses (or government) had offices. Where

[172] 'Food Sovereignty Ghana and COFAM demonstrates against GMO's', *Your Neighborhood*, May 12th–May 26th, 2014, sec. 1.

[173] Owutse Noah (student, Department of Family and Consumer Sciences, University of Ghana, Legon), in discussion with the author, 27 May 2014.

[174] Agbeve Worlali, interview by author.

Fig. 7.10 Salad with macaroni & tinned beans, Aburi Gardens

tourists were likely to visit, they were served continental (or English) foods. Osseo-Asare contended that this behaviour reflected hybridity whereby the development of a national cuisine was a post-colonial process of vernacularisation, instead of a process of cultural domination (Fig. 7.10).[175]

In June of 2016, "Susie", a programme officer at the Ghana Association of Food Producers, was interviewed by Joeva Rock, an anthropologist from New York University. Susie was protesting the Gates and Rockefeller Foundations donation of funds for the adoption of Genetically Modified Organisms (GMOs) across Africa. In response to the Gates Foundations assertion that GMOs are the answer to African hunger, Susie stated: "How do you know that is the answer to our hunger? And even who said we are hungry? We are not even starving. Eh? We are not starving."[176] Indeed, in my own observations of major markets, cold storage businesses, grocery stores, roadside food vendors, ports, restaurants, hotels,

[175] Osseo-Asare, interview by author.

[176] Joeva Rock, '"We Are Not Starving": Challenging Genetically Modified Seeds and Development in Ghana', *Culture, Agriculture, Food and Environment* 41, no. 1 (June 2019): 15–23.

resorts, and various interviews with workers at government ministries and NGOs inadequate pre-market storage was the problem as refrigeration was virtually non-existent due to frequent power outages. Rather than a lack of food, there was plenty of food available, some of it rotting in the markets for want of improved transportation and storage.[177]

Since the colonial encounter, the difficulty of intra-regional transport of food continues to be an issue. Edwin Kweku Andoh Baffour, communications head for FSG—an organisation that lobbies against NGOs that introduce Monsanto foods into Ghana—confirmed this in a press conference after a court case that contested the legitimacy of genetically-modified rice and cowpea research in Ghana: "We produce more food than we need. Food rots. What we rather need is investment in roads, warehouses, [to] manage post-harvest loss."[178] What organisations like FSG fear is the successful adoption of GMO foods in Ghana would encourage the growth of the national deficit. GMO foods need artificial fertiliser, something that is too expensive for most farmers and would require government subsidies and thus, greater government outlay. Investment in transport networks that facilitate the conveyance of food into urban areas could help to lower prices for both local and imported foods. At present, the poor intraregional transport network and lack of suitable warehouse space contributes to the increase of frozen chicken imported from Europe, the US, and Brazil.[179]

Aside from transport difficulties, the other constant from the twentieth century to the present is the continued production of regional cuisines. Regional cuisines are now an accepted part of the national identity and have become a recognised and celebrated aspect of Ghanaian culture. Commensality practices, whether on the coast, the central forested regions, or in the north, have changed little from one century to the next. The long-documented practice of individuals taking meals separately from family does not inhibit the unity of family. Indeed, it is the act of

[177] Miller, 'Food and Nationalism', 73.

[178] Rock, 'We Are Not Starving'.

[179] Bettina Rudloff and Evita Schmieg, 'European Chicken Drumsticks for West Africa: A Threat to Local Markets?', *Rural 21 The International Journal for Rural Development* 54, no. 2 (2019); Chris Ward, 'EU Chicken Dumping Starves Africa', *Mail and Guardian*, 10 November 2017, sec. Business; Petra Titze et al., 'No More Chicken, Please', *Church Development Service*, 2018.

cooking of regional foods that demonstrates care and connection in the community.[180]

Another enduring food practice, that of urban farming, contributes toward the reproduction of local cuisines. Urban farming serves to maintain cultural food practices while at the same time providing income for men and women who participate in the informal economy.[181] Much of the food grown in these areas historically were and continue to be "exotics"—European vegetables like peppers, beetroots, cauliflower, and herbs. Indigenous vegetables, including okra, eggplant, and leafy greens, are staples for the urban gardener and remain critical to the cultural reproduction of local dishes.[182] Foods are still imported, such as plantain from Abidjan and tomatoes from Burkina Faso, in order to meet robust demand in urban areas (Fig. 7.11).[183]

Conclusion

Despite the post-Second World War food fight that launched its bid for independence, Ghana's story demonstrates the centrality of local foods and cooking to its national character. The cultural weight of regional traditional foods, their power to satisfy, and the overall collective social emphasis on the main meal of the day have persisted in Ghana, irrespective of centuries of trade with Europeans. Food intersects with gender, authority, and ethnicity in Ghana and the proper meal reveals much about one's regional identity and the social role one occupies in relation to that identity. Traditional ways of cooking and eating changed but not by much, irrespective of the colonial experience as imports became necessary staples, and despite both colonial and post-independence efforts by the government to limit their use. These foods were either absorbed into

---

180 Helena Tuomainen, 'Eating alone or together?', 42; Psyche Williams-Forson, 'I Haven't Eaten', 74.

181 Asafu-Adjaye, 'The Tendency to Urban-Farm in Accra', 165.

182 Asomani-Boateng, 'Urban Cultivation in Accra', 592.

183 'It Is a National Scandal for Ghana to Import Food—Akufo-Addo', accessed 25 July 2019, https://www.ghanaweb.com/GhanaHomePage/NewsArchive/It-is-a-national-scandal-for-Ghana-to-import-Food-Akufo-Addo-509737.

Fig. 7.11 "European produce" at Citiveg, U of Ghan

the proper main meal of the day or used to fill the other mealtimes of the day with quick, convenient foods.

There was (and still is) no sign of the conscious development of a national cuisine in Ghana. Locally produced food and regional dishes continue to create communities and to define boundaries. There are other countries in the region with similar structural and environmental strictures. Several have adopted the cuisine of their colonising country as part of their national identity, such as Equatorial Guinea, South Africa, and Gabon.[184] However, in large part due to structural factors that exacerbated regional ways of cooking and eating, the adoption of a national cuisine has been a non-starter in Ghana. In fact, in recent years, the regional character of Ghanaian food has been embraced by the citizenry as a national good, and as something that makes them unique among nations.[185]

[184] Cusack, 'African Cuisines', 214; Nugent, 'Do Nations Have Stomachs?', 107.
[185] Simpson Miller, 'Food and Nationalism', 2.

References

Accra, Director of Supplies, Tinned beef order from Australia. CSO 8/1/230. PRAAD Accra, 7 April 1942

M. Adinkrah, *Witchcraft, Witches, and Violence in Ghana* (Berghahn Books, 2015)

F. Agbodeka (ed.), *Handbook of Eweland: The Ewes of Southeastern Ghana*, vol 1 (Woeli Publishing Services, Accra, 1997)

Agricultural Mutual Improvement Association, Gold Coast Nation, 7 April 1917

K.O. Akurang-Parry, "Untold difficulties:" the indigenous press and the economic effects of the first world war on Africans in the gold coast, 1914–1918. Afr. Econ. Hist. **34**, 45–68 (2006). https://doi.org/10.2307/25427026

E.K. Akyeampong, *Drink, Power, and Cultural Change* (Heinemann, Portsmouth, 1996)

E.K. Akyeampong, Race, identity and citizenship in Black Africa: the case of the Lebanese in Ghana. Africa (London. 1928) **76**(3), 297–323 (2006). https://doi.org/10.3366/afr.2006.0033

Alata, Osu Mantse, and Oyokohene, 14 February 1948 Boycott Speech. ARG 1/1/254. PRAAD Accra, n.d.

L. Alfers, R. Abban, Occupational health & safety for indigenous caterers in Accra, Ghana. Occup. Health **24**, p. 5 (2011)

J.M. Allman, V.B. Tashjian, *'I Will Not Eat Stone': A Women's History of Colonial Asante* (James Currey, Portsmouth, 2000)

J. Amikuzuno, Border effects on spatial price transmission between fresh tomato markets in Ghana and Burkina-Faso: any case for promoting trans-border trade in West Africa? In *85th Annual Conference*. Warwick University, Coventry, UK, 2011. https://ageconsearch.umn.edu/record/108943

K. Arhin, *The Life and Work of Kwame Nkrumah: Papers of a Symposium Organized by the Institute of African Studies, University of Ghana, Legon* (Africa World Press, Trenton, 1993)

P. Asafu-Adjaye, The tendency to urban-farm in Accra: a cultural lag-labor surplus Nexus. J. Third World Stud. **29**(2), 159–182 (Fall 2012)

R. Asomani-Boateng, Urban cultivation in Accra: an examination of the nature, practices, problems, potentials and urban planning implications. Habit. Int., Institutional Analysis of Asia-Pacific Real Estate Markets **26**(4), 591–607 (1 December 2002)

M.A. Awo, Marketing and market queens: a case of tomato farmers in the upper east region of Ghana. Rheinischen Friedrich-Wilhelms-Universität zu Bonn, 2010

M.A. Awo, *Marketing and Market Queens: A Study of Tomato Farmers in the Upper East Region of Ghana* (LIT Verlag Münster, Berlin, 2012)

D.A. Ayensu, *The Art of West African Cooking* (Doubleday, Garden City, 1972)

B&G Foods, Inc, Ac'cent flavor enhancer. Accent Flavor. https://www.accentflavor.com/about. Accessed 2 July 2019

Bank to grant loans to sugar-cane farmers. The Ghanaian Times. PRAAD Accra, 17 December 1968

J. Bossalini, Director, Fishing industry in the gold coast by Messrs. John Netherwood And Co., London. CSO 8/21/2. PRAAD Accra, 18 April 1933

Brong Ahafo Annual Report. Accra, Ghana: PRAAD, Accra. ARG 1/17/24, 1919

L. Brydon, K. Legge, *Adjusting Society: The World Bank, the IMF, and Ghana* (I.B.Tauris, London, 1996)

R.F. Burton, *Wanderings in West Africa from Liverpool to Fernando Po* (Tinsley brothers, London, 1863)

K.A. Busia, Dr., Sekondi-Takoradi social survey—school feeding. ADM 23/1/2768. PRAAD Accra, 9 December 1947

A.W. Cardinall, *The Gold Coast, 1931* (Government Printer, Accra, 1932)

J.A. Carney, *Black Rice: The African Origins of Rice Cultivation in the Americas,* New Ed edition (Cambridge, MA: Harvard University Press, 2002)

Central Advisory Committee on Education, Meals for Schoolchildren—minutes. ADM 23/1/2768. PRAAD Accra, 12 February 1946

E.A. Charles, *Precolonial Senegal: The Jolof Kingdom, 1800–1890* (Boston University, African Studies Center, 1977) https://open.bu.edu/handle/2144/23142

Chief Commissioner's Office, News Officer, Fun fair in aid of war charities. ARG 1/1/215. PRAAD Kumasi, 19 February 1942

G. Clark, Money, sex and cooking: manipulation of the paid/unpaid boundary by Asante market women, in *The Social Economy of Consumption*, ed. by H. J. Rutz, B. S. Orlove, (University Press of America, Lanham, MD, 1989)

G. Clark, *Onions are My Husband: Survival and Accumulation by West African Market Women* (University of Chicago Press, Chicago, 1994)

G. Clark, From fasting to fast food in Kumasi, Ghana, in *Food Consumption in Global Perspective: Essays in the Anthropology of Food in Honour of Jack Goody*, ed. by J. Klein, A. Murcott, (Palgrave Macmillan, Basingstoke, 2014)

L. Clifford (ed.), *Our Days on the Gold Coast* (John Murray, London, 1919)

E.J. Collins, Ghanaian highlife. Afr. Arts **10**(1), 62–68 (1976)

Colonial Economic Policy, (Hansard), 17 December 1940., https://api.parliament.uk/historic-hansard/lords/1940/dec/17/colonial-economic-policy. Accessed 17 March 2020

Colonial Government, Gold Coast, Fish curing development. ADM 23/1/401. PRAAD Accra, 1921

Commissioner, District, Sunyani district commissioner's letter to Kumasi chief commissioner on confiscated goods. ARG 1/1/221. PRAAD Kumasi, 19 December 1944

Comptroller of Customs, Memo on meat imports from the UK. CSO 8/1/229. PRAAD Accra, 25 March 1941

F. Cooper, *Africa since 1940: The Past of the Present* (Cambridge University Press, Cambridge, 2002)

I. Cusack, African cuisines: recipes for nation-building? J. Afr. Cult. Stud. **13**(2), 207–225 (1 December 2000)

P. De Marees, *Description and Historical Account of the Gold Kingdom of Guinea* (British Academy, Oxford, 1987)

A. Dede, *Ghanaian Favourite Dishes* (Anowuo Educational Publications, Accra, 1969)

J. Dieterle, Basel Mission Report Aburi. Aburi, Ghana: PRAAD, Aburi. ABM, D-1,9, 1858

District Commissioner Juaso, Permission to rail foodstuffs. ARG 6/12/7. PRAAD Kumasi, 17 September 1945

Documents Relating to the First Session of the Food And Agriculture Conference of the United Nations, Quebec, Canada, 16 November 1945. MSS. Brit. Emp. S. 332 1944–1948. Bodleian Library Special Collections

S. Donkoh, M. Tachega, N. Amowine, Estimating technical efficiency of tomato production in Northern Ghana. Am. J. Exp. Agri. **3**, 56–75 (1 December 2012). https://doi.org/10.9734/AJEA/2013/1447

Duncan-Johnstone, Captain A.C, Minutes of the meeting held in the director of supply's office. Meeting Minutes. Accra: Chief Commissioner Ashanti—Gold Coast Colony. ARG 1/1/67. PRAAD Accra, 25 October 1941

Dunkor, Staff Welfare Officer, Alfred Moses. The Kumasi hospital canteen. ARG 13/1/11. PRAAD Kumasi, 13 June 1952

F. Efua Dovlo, Nutrition and food utilization, in *Handbook of Eweland: The Ewes of Southeastern Ghana*, ed. by F. Agbodeka, vol. 1, (Woeli Publishing Services, Accra, 1997)

J. Fabian, Popular culture in Africa. Africa **48**(4), 315–334 (1978)

M.J. Field, *Gold Coast Food* (Achimoto College Press, Achimoto, 1931)

M.J. Field, Gold coast food. Petits Propos Culinaire **43**, 7–19 (1994)

Free supply of sugar at trade fair. The Ghanaian Times. PRAAD Accra. 10 October 1966

K. Frimpon, *The Boy in the Oversized Smock: School Memories in Living Color* (Xlibris Corporation, 2010)

From Lebanon to Africa., https://www.aljazeera.com/program/al-jazeera-world/2015/10/28/from-lebanon-to-africa/. Accessed 29 September 2020

B.Y. Gebe, Ghana's foreign policy at independence and implications for the 1966 Coup D'état. J. Pan Afr. Stud. **2**(3), 28 (2008)

R. B. S. George (ed.), *Material Life In America, 1600–1860* (Northeastern University Press, Boston, n.d.)

GhanaCelebrities.Com, The wonderful Ghanaian CHOP BAR Experience, 25 September 2014., https://www.ghanacelebrities.com/2014/09/25/wonderful-ghanaian-chop-bar-experience-tried/

Gold Coast Government, Report on the census. ADM 5/2/5. PRAAD Accra, 1921

Gold Coast Marketing Officer, Mr. Packham, Consignment of cabbage received from dept. of agriculture, Nigeria. CSO 8/4/8. PRAAD Accra, 10 March 1941

J. Goody, *Cooking, Cuisine and Class: A Study in Comparative Sociology* (Cambridge University Press, Cambridge, 1982)

J. Goody, E.N. Goody, Food and identities: changing patterns of consumption in Ghana. Camb. Anthropol. **18**(3), 1–14 (1995)

Great Britain and Commission on Marketing of West African Cocoa, *Report of the Commission on the Marketing of West African Cocoa* (H. M. Stationery Off, London, 1938)

S.E. Greene, Developing the arts for development: perspectives on Ghana. Afr. Notes, 1–8 (1998)

Grow more flags. The Daily Echo, 12 May 1954

D.M. Guggisberg, *We Two in West Africa* (W. Heinemann, London, 1909)

I.K. Gyasi, Aspects of English in Ghana. English Today 7(2), 26–31 (April 1991). https://doi.org/10.1017/S0266078400005502

J.R. Ham, Cooking to be modern but eating to be healthy: the role of Dawa-Dawa in contemporary Ghanaian foodways. Food, Cult. Soc. **20**(2), 237–256 (2017). https://doi.org/10.1080/15528014.2017.1305827

J. Hess, Exhibiting Ghana: display, documentary, and "national" art in the Nkrumah era. Afr. Stud. Rev. **44**(1), 59–77 (2001)

Hints for Christmas cooking. The Ghanaian Times. PRAAD Accra, 24 December 1968

Interview with Dr. Kofi Amoah. State of Affairs. Ghana: GHOne, 2 December 2016

F.R. Irvine, *Woody Plants of Ghana: With Special Reference to Their Uses* (Oxford University Press, Oxford, 1961)

It is a national scandal for Ghana to import food—Akufo-Addo. https://www.ghanaweb.com/GhanaHomePage/NewsArchive/It-is-a-national-scandal-for-Ghana-to-import-Food-Akufo-Addo-509737. Accessed 25 July 2019

S. Jallow, A feasibility study into the establishment of a Gambian restaurant in Accra, 2018., https://air.ashesi.edu.gh/handle/20.500.11988/392

A.T. Knight, Draft order to restrict railing of local staples from Konogo or Juaso railway. ARG 6/12/7. PRAAD Kumasi, 1939

J. Knörr, C. Kohl, *The Upper Guinea Coast in Global Perspective* (Berghahn Books, 2016)

S. Kolavalli, M. Vigneri, Cocoa in Ghana: shaping the success of an economy, in *Yes, Africa Can: Success Stories from a Dynamic Continent*, ed. by P. Chuhan-Pole, M. Angwafo, (World Bank Group Publications, Washington, D.C, 2011)

R. Korley, Application to run canteen at Central Hospital, Kumasi. ARG 13/1/11. PRAAD Kumasi, 22 January 1957

J. Kraus, The struggle over structural adjustment in Ghana. Afr. Today **38**(4), 19–37 (1991)

Kumasihene, By-laws concerning consumption of liquor at funeral customs. ARG 26/8/12. PRAAD Accra, 1928

T.E. Kyei, *Our Days Dwindle: Memories of My Childhood Days in Asante*. Edited by Jean Allman (Portsmouth: Heinemann Educational Books, 2001)

J.D. La Fleur, *Fusion Foodways of Africa's Gold Coast in the Atlantic Era* (Brill, Leiden, 2012)

A. Logan, A history of food without history: food, trade, and environment in West-Central Ghana in the second millennium AD. PhD diss., University of Michigan, 2013

F. Lyon, Trust, networks and norms: the creation of social capital in agricultural economies in Ghana. World Dev. **28**(4), 663–681 (2000)

S.O. Marfo, Immigrant businesss in Ghana: a study of the Lebanese in Accra. Thesis, University of Ghana, 2012., http://ugspace.ug.edu.gh/handle/123456789/5919

J.C. McCann, *Stirring the Pot: A History of African Cuisine* (Ohio University Press, Athens, Ohio, 2009)

M. Meredith, *The State of Africa: A History of the Continent Since Independence* (Simon & Schuster, S.l., 2013)

S.F. Miescher, *Making Men in Ghana* (Indiana University Press, Bloomington, 2005)

B. Miller, Food and nationalism in an Independent Ghana. MA Thesis, Georgia State University, 2015

J. Milne, *Kwame Nkrumah, a Biography* (Panaf Books, London, 2006)

Ms. Ama Akoto personal interview with Brandi Simpson Miller at Ghana Tourism Ministry, Accra, 28 May 2014

National Commission on Culture, The cultural policy of Ghana. Ghanaian Government, 2004., http://www.ghanaiandiaspora.com/wp/wp-content/uploads/2014/09/CULTURAL.-POLICY-FINAL.pdf

Nigeria Jollof Delicious—Mark Zuckerberg. https://www.ghanaweb.com/GhanaHomePage/entertainment/Nigeria-jollof-delicious-Mark-Zuckerberg-466952. Accessed 28 April 2020

K. Nkrumah, Circular from Prime Minister Kwame Nkrumah to Chief Regional Officers on Behalf of The National Food Board. ARG 2/8/27. PRAAD Accra, 1953

Northern Territories Annual Report with Descriptions of Local Produce and Revenue. Accra, Ghana: PRAAD, Accra. ARG 1/17/24, 1919

P. Nugent, Do nations have stomachs? food, drink and imagined community in Africa. Afr. Spectr. **45**(3), 87–113 (28 March 2011)

Office of the Government Statistician, *Accra Survey of Household Budgets* (Government Printing Department, Accra, 1953)

D.C. Ohadike, The influenza pandemic of 1918–19 and the spread of Cassava cultivation on the lower Niger: a study in historical linkages. J. Afr. Hist. **22**(3), 379–391 (July 1981)

J.D. Osei-Tutu, Dr., Personal interview with Brandi Simpson Miller at University of Legon, History Department, 27 May 2014

F. Osseo-Asare, "We eat first with our eyes": on Ghanaian cuisine. Gastronomica **2**(1), 49–57 (2002)

F. Osseo-Asare, *Food Culture in Sub-Saharan Africa* (Greenwood Press, Westport, 2005)

F. Osseo-Asare, Interview by author, 29 March 2013

F. Osseo-Asare, B. Baëta, *The Ghana Cookbook* (Hippocrene Books, New York, 2015)

Our sugar problem can be solved if…. The Ghanaian Times. PRAAD Accra, 18 October 1968

Papers of Sir Charles Henry Harper 1914–1920. Oxford. MSS. Brit. Emp. S. 344, Box 2. Bodleian Library Department of Special Collections, 1 June 1920

D.W. Phillipson, *African Archaeology* (Cambridge University Press, 2005)

J.M. Pilcher, *Que Vivan Los Tamales!: Food and the Making of Mexican Identity* (University of New Mexico Press, Albuquerque, 1998)

Police seize 600 cartons of sugar. The Ghanaian Times. PRAAD Accra, 17 October 1968

F.M. Purcell, Nutrition in the colonial empire. BMJ **2**(4100), 294–296 (1939)

O. Quist-Arcton, Jollof Rice: West Africans dish it up with a Hefty serving of smack talk. NPR.org, https://www.npr.org/sections/thesalt/2016/08/30/491380208/jollof-rice-west-africans-dish-it-up-with-a-hefty-serving-of-smack-talk?t=1568145970707. Accessed 10 September 2019

Recommendations with regard to nutrition in the colonial territories. Papers of Arthur Creech Jones. Oxford, March 1944. MSS. Brit. Emp. S. 332 1944–1948. Bodleian Library Special Collections

J. Rich, *A Workman Is Worthy of His Meat: Food and Colonialism in the Gabon Estuary* (University of Nebraska Press, Lincoln, 2007)

C.C. Robertson, *Sharing the Same Bowl?: A Socioeconomic History of Women and Class in Accra, Ghana* (Indiana University Press, Bloomington, 1984)

J.E. Robins, "Food comes first": the development of colonial nutritional policy in Ghana, 1900–1950. Glob. Food Hist. **4**(2), 168–188 (2018)

J. Rock, "We are not starving": challenging genetically modified seeds and development in Ghana. Cult. Agri. Food Environ. **41**(1), 15–23 (June 2019)

G. J. Rodger, Memorandum on Sierra Leone Rice. ADM 11/1/89. PRAAD Accra, 16 December 1909

B. Rudloff, E. Schmieg, European chicken drumsticks for West Africa: a threat to local markets? Rural 21. Int. J. Rural Dev. **54**(2) (2019) https://www.rural21.com/english/search/detail/article/european-chicken-drumsticks-for-west-africa-a-threat-to-local-markets.html

S.J. Salm, T. Falola, *Culture and Customs of Ghana* (Greenwood Press, Westport, Conn, 2002) http://archive.org/details/culturecustomsof00salm

Secretary of State for the Colonies, Telegram on Argentine tinned butter supplies. PRAAD Accra, 18 June 1945

G. Serra, Towards a political economy of statistics: a study of household budget surveys in the gold coast, 1945–1957, in *Session on Measurement, Planning and the State in Sub-Saharan Africa: Historical Perspective*, vol. 33, (Simon Fraser University, Vancouver, 2013)

U. Sill, *Encounters in Quest of Christian Womanhood* (Brill, Leiden, 2010)

S. Stevano, D. Johnston, Childrens food consumption away from home and food provisioning in Accra. Presented at the SOAS Food Studies Centre, SOAS, University of London, 27 January 2017

A. Subedar, I. Ahmed, West Africa steams over Jollof Rice War. BBC News, 26 August 2017., https://www.bbc.co.uk/news/blogs-trending-41053424

J.A. Tambo, T. Gbemu, Resource-use efficiency in tomato production in the Dangme West District, Ghana. In *International Research on Food Security, Natural Resource Management and Rural Development*, 5. Zurich, 2010., http://www.tropentag.de/2010/proceedings/node2.html

The Gold Coast Cookery Book (Accra: British Red Cross, 1933)

E.K.W. Thompson, Inspector general of police to commissioner of police Ashanti. ARG 1/1/138. PRAAD Kumasi, 1928

P. Titze, J. Rode, F. Mari, R. Lange, No more chicken, please. Church Development Service, 2018., https://library.concordeurope.org/record/2020/files/DEEEP-PAPER-2018-032.pdf

H. Tuomainen, Eating alone or together? Commensality among Ghanaians in London. Anthropol. Food **14**, S10 (2014). https://doi.org/10.4000/aof.7718

A. Tweedie, *Melting Pot: Ghanaian Cooking with a Twist* (Asantewaa Publication, Accra, 2011)

Wagon, Butter. CSO 8/8/1. PRAAD Accra, 1939

Wagon, Telegram referring to milk and wheat supplies from Canada. CSO 8/1/233. PRAAD Accra, 6 February 1942

C. Ward. EU chicken dumping starves Africa. Mail and Guardian, 10 November 2017, sec. Business, https://mg.co.za/article/2017-11-10-00-eu-chicken-dumping-starves-africa/

When you get visitors…. The Ghanaian Times. PRAAD Accra, 7 October 1966

R. Wilk, *Home Cooking in the Global Village: Caribbean Food from Buccaneers to Ecotourists* (Berg, Oxford, 2006)

P. Williams-Forson, "I haven't eaten if i don't have my soup and Fufu": cultural preservation through food and foodways among Ghanaian migrants in the United States. Afr. Today **61**(1), 68–87 (2014)

Worlali, Agbeve personal interviewed by Brandi Miller in Legon, 27 May 2014

I.E.A. Yeboah, Structural adjustment and emerging urban form in Accra, Ghana. Afr. Today **47**(2), 61–89 (2000)

J.O. Yorke, E.K. Amissah, An overview of Kwame Nkrumah's cultural policies on Ghana's visual culture. Res. J. Humanit. Cult. Stud. **3**(3), 12 (2017)

S. Zubaida, Drink, meals and social boundaries, in *Food Consumption in Global Perspective: Essays in the Anthropology of Food in Honour of Jack Goody*, ed. by J. Klein, A. Murcott, (Palgrave Macmillan, Houndmills, 2014)

8

Globalisation and Local Foodways in Ghana

Globalisation, defined as 'a process of greater integration within the world economy through movements of goods and services, capital, technology and (to a lesser extent) labour,' is affecting the way food is marketed in Ghana today.[1] In 2019 Madam Evangeli, proprietor of OJ's Kitchen in East Legon Accra, was encouraged by one of her customers to list herself as a vendor on the platform Jumia Food Portal. At that time Madame Evangeli had built herself a solid reputation as a red-red seller and relied on walk up customers for most of her business.[2] After the COVID 19 pandemic began, the technology transformed her life. By December of that year, she began to get so much business from the Jumia Food Portal that she had to expand her stall area and to hire on several employees to keep up with demand. Since the start of the pandemic, she has expanded into catering and has added a rice combo as well as breakfast sausage and eggs to her menu.

[1] Rhys Jenkins, 'Globalization, production, employment and poverty: debates and evidence,' *Journal of International Development* 16, no. 1 (2004): 1, https://doi.org/10.1002/jid.1059.

[2] Adwoa Adubia, 'Changing lives through the internet; The story of a Ghanaian beans seller,' Thesunghana.com, last accessed May 11, 2021, https://www.thesunghana.com/2021/04/07/changing-lives-through-the-internet-the-story-of-a-ghanaian-beans-seller/.

B. Simpson Miller, *Food and Identity in Nineteenth and Twentieth Century Ghana*, Food and Identity in a Globalising World,
https://doi.org/10.1007/978-3-030-88403-1_8

Shortly after Madam Evangeli experienced an ecommerce revolution in her business of selling both local and 'glocal' (meant here as local fast-food restaurants mimicking global chains and western style meals), Pizza Hut—run by an Indian conglomerate—opened two branches: one in Osu, Accra on Oxford Street, and another in East Legon. They offer carry out, car side pickup, and COVID safe contactless delivery—all via their online ordering and pizza tracking technology.[3] New technology, partially ushered into more common use by the COVID pandemic, is used to market both traditional and 'fast food' and marks the latest impact of globalisation on Ghanaian foodways.

These examples demonstrate that today, as in the past, Ghanaians continue to experience hybridization in the borrowing from and incorporation of foreign foods and, in this case, technologies. There are three generalized responses to globalisation of foodways: *homogenization*—where global culture is becoming standardized around western/American patterns, *polarization*—where there is resistance to western cultural norms, or *hybridization*—where elements are borrowed and incorporated.[4] While foreign foods are being integrated into the main meal of the day and into other mealtimes, polarization in reaction to changing ways of eating and of food production is seen in growing concerns about nutritional health. Discourses on lifestyle diseases like obesity and diabetes, are increasingly common. Additionally, anxiety over the convenience and impersonalisation of food manifests itself in food safety concerns and disquiet about the loss of regional foodways, especially in urban areas. Finally, apprehension over threats to local control of food production and fears over the loss of Ghanaian sovereignty recall historical issues with transport infrastructure that challenge Ghana's ability to maximise its food production, giving rise to economic vulnerability. This chapter explores globalisation's effects on the foodways in Ghana over the last ten

[3] Shahaduz Zaman, Nasima Selim, and Taufique Joarder, 'McDonaldization without a McDonald's', *Food, Culture & Society* 16, no. 4 (1 December 2013): 560, https://doi.org/10.2752/175174413X13758634982010.; 'Domino's Pizza, world's largest pizza company, set to launch in Accra,' Ghana Web, last accessed May 11, 2021, https://www.ghanaweb.com/GhanaHomePage/NewsArchive/Domino-s-Pizza-world-s-largest-pizza-company-set-to-launch-in-Accra-1211608?gallery=1.

[4] Robert Holton, 'Globalizations Cultural Consequences,' *The Annals of the American Academy of Political and Social Science*, 570 (2000): 141.

years by considering current concerns over the nutritional health of its citizens that packaged and fast foods represent, challenges to national security that reliance on imported foods constitutes and concludes with the creative ways that Ghanaians are responding to these challenges.

In recent discussions of how globalisation affects Ghanaian foodways, a controversial issue has been whether globalisation has had a detrimental effect upon nutrition and health of Ghanaians. Scholars argue that urbanization has had the biggest impact on diets as people are looking for time saving shortcuts in packaged foods and convenient imports like wheat and rice.[5] Scholar Fran Osseo-Asare commented that people in Ghana are looking for labour saving shortcuts like Maggi seasoning cubes versus free range meats and beans. She stated that wheat consumption was up 170% since 2005, attributable to the USDA and the Canadian government supplying Ghana with their surplus wheat.[6] Indeed, since 2015 to 2019 imports of grains like rice have increased by 85%.[7] According to Osseo-Asare, wheat flour is popular due to its portability and convenience. She ascribes rising rates of diabetes to the increase in consumption of wheat.

Diabetes is not the only illness associated with an increased consumption of imported foods. In his February 2020 State of the Nation Address the president of Ghana, Nana Addo Dankwa Akufo-Addo, remarked on the worrying trend of obesity in the country. "The food we eat, the mode of cooking and lack of exercise are all having a great impression on our health, we need to take responsibility of our individual health, our health is very much determined by our lifestyle."[8] Obesity has been linked to an increased consumption of fast food, which is growing at an annual rate of

---

[5] Jessica Ham (Assistant Professor of Anthropology, Oxford College, Emory University), interviewed by Brandi Simpson Miller, via telephone, April 21, 2021, Abingdon, Oxfordshire, UK.

[6] Barbara Baeta (Proprietor, Flair Catering, Cantonments, Accra, Ghana), 'Brandi Miller's requested interview,' interviewed by Fran Osseo-Asare, Flair Catering, Cantonments, Accra, March 22, 2018.

[7] 'Country Briefs—Ghana,' GIEWS—Global Information and Early Warning System, FAO, accessed May 2021, http://www.fao.org/giews/countrybrief/country.jsp?code=GHA.

[8] 'Akufo-Addo concerned about Ghana's obesity rate,' Ghana News, myjoyonline.com, accessed May 2021, https://www.myjoyonline.com/akufo-addo-concerned-about-ghanas-obesity-rate/; 'SONA 2020: Akufo-Addo expresses worry about obesity in Ghana,' Adom Online, accessed May 2021, https://www.adomonline.com/sona-2020-akufo-addo-expresses-worry-about-obesity-in-ghana-video/?fbclid=IwAR11jb2CQTa-6_aoxU_JV-AqcKGj8bzkt-4LO-bC1kBR02-8iiUo6ejeZ5I.

20%.[9] Most of what is purchased—fried rice, jollof rice, and fried chicken—is being consumed by singletons, middle and upper class people, and young people. Barbara Baëta, cookbook publisher, catering school proprietor, and official state caterer, corroborated this data in her observations of daily life in Accra. She stated that Asian influences can be seen in stir fries made from 'polished grains' that are sold on the street. She worried about the 'quick meals kids are getting' that are filled with MSG from Maggi cubes. She observed that the adverts for these ready foods are 'fantastic', and these ads contributed to increased consumption of pre-packaged foods by young people who are suffering the nutritional consequences.[10]

Baëta is correct in her observation that adverts influence the eating habits of young people in urban areas. Upwards of 70% of students appear to have a high degree of autonomy with respect to food choices, especially if they walk or take public transportation to school. These children encounter Fan Milk or Milo (Nestle) vendors on the streets who in the case of Nestle, also sponsor nutritional education kits.[11] Socioeconomic inequalities and the availability of inexpensive prepared foods are embedded in the urban organization of the environment, and in the larger globalized food systems. The present rate of childhood obesity in cities like Greater Accra and Kumasi is between 8.6 and 10.7%. The paradox of this obesity is that children are suffering from undernutrition even though they are consuming an excess of calories.[12] It is not just financially disadvantaged children who are suffering from an increased availability of prepacked foods. The children of affluent parents (characterized by mothers with a post-secondary education) have access to more processed foods at

---

[9] James Boafo, 'Ghanaians are eating more fast food: the who and the why,' The Conversation, accessed May 2021, https://theconversation.com/ghanaians-are-eating-more-fast-food-the-who-and-the-why-153810.

[10] Barbara Baeta, interview.

[11] Sara Stevano, Deborah Johnston, Emmanuel Codjoe, Better decisions for food security? Critical reflections on the economics of food choice and decision-making in development economics, *Cambridge Journal of Economics*, Volume 44, Issue 4, July 2020, Pages 828, https://doi.org/10.1093/cje/beaa012.

[12] Prince Kwaku Akowuah, Emmanuel Kobia-Acquah, 'Childhood Obesity and Overweight in Ghana: A Systematic Review and Meta-Analysis,' *Journal of Nutrition and Metabolism* (2020), https://doi.org/10.1155/2020/1907416.

home. These affluent children also tend to watch on average two or more hours a day of TV. Additionally, even though the Ministry of Health in Ghana recommends 1–2 hours of moderate physical activity each day, it is usually only children who show potential to be competitive athletes who get the opportunity to engage in school-based physical activity.[13]

It is not just the lack of exercise or increased access to packaged foods that people find concerning. What disturbs some is that packaged foods encourage eating outside of traditional agricultural rhythms. Stevano et al found that less affluent children receiving low or irregular chop money results in children having their first meal at the first or second break of the school day.[14] Eating a first or second meal of the day later in the morning was characterized as 'experiencing food insecurity'. This behaviour may not be due to food insecurity, however. Some locals would say that 'eating has its time'. Fasting was an historical reality before refrigeration and other methods of preservation became available. '…if we eat carelessly and lawlessly, it can cause sicknesses and diseases and cut short our life expectancy. We become sick and die earlier than we are supposed to live if we eat anyhow and at any time.'[15] The daily rhythms of rural folks who go to work on their farms without breakfast often do not eat their first meal until noon are held up as the traditional ideal and best for a long and healthy life.

It is not just the rhythms of eating, but food safety concerns over the packaging of traditional fast foods like banku—now often wrapped in polyethene instead of a leaf or maize husk—that are causing feelings of unease. Kofi Essel, Head of the Ghanaian Food and Drugs Authority stated that eating food wrapped in polythene regularly can cause cancer in the stomach because particles from the plastic that can get into the

---

[13] Aryeetey R, Lartey A, Marquis GS, Nti H, Colecraft E, Brown P. 'Prevalence and predictors of overweight and obesity among school-aged children in urban Ghana,' *BMC Obesity 4*, no. 4 (2017): 6, https://pubmed.ncbi.nlm.nih.gov/29214030/.

[14] Sara Stevano, Deborah Johnston, and Emmanuel Codjoe, 'The Urban Food Question in the Context of Inequality and Dietary Change: A Study of Schoolchildren in Accra,' *The Journal of Development Studies* 56, no. 6 (2 June 2020): 10, https://doi.org/10.1080/00220388.2019.1632434.

[15] Edward Bamfo-Darko, 'Your Eating Pattern Determines Your Health And Lifespan. Including Should Women Fast?,' accessed May 2021, https://www.modernghana.com/news/1073413/your-eating-pattern-determines-your-health-and.html.

food.[16] Food safety concerns also extend to other glocal fast foods sold on the street like jollof rice, fried rice, or fried plantain. Food consumption away from home is an increasing phenomenon among all subcategories of the population and concerns about food hazards such as pesticide residue in vegetables or excessive use of artificial flavour enhancers are growing concerns for consumers in Ghana.[17]

Notwithstanding these concerns over nutrition and food safety, others argue that the most important food issues in Ghana are related to the lack of promotion of regional cuisines. In the words of one of this view's main proponents, urbanization has a huge impact on diets in Ghana, contributing to a huge increase in packaged foods. However, the sense is that people *want to be* eating traditional food for various reasons.[18] According to this view the consumption and promotion of regional foods does triple duty—regional cuisines are closely tied to cultural identity, with the preservation of the environment, and the protection and bolstering of the national economy. Take for example The Ghana Food Movement (GFM). GFM started in response to the increasing lack of regional cuisines and ingredients found in urban centres. This organization was founded upon the belief that promoting local food production is good for the economy, the environment and for the preservation of local identities.[19] GFM seeks to generate dialogue about the lack of diversity in urban diets by running an online platform and promoting offline meetups to eat delicious local food.

The growing promotion of regional foods is echoed in the sentiments of people who commented on the recent opening of the Domino's Pizza branches. Some of these comments took on distinctly patriotic overtones. On 22 March of 2021, one commenter wrote 'The Ghanaian FDA Must Wake Up':

---

[16] 'Polythene wrapped banku: a more deadly substitute for cigarettes?,' accessed May 2021, https://www.ghanaweb.com/GhanaHomePage/NewsArchive/Polythene-wrapped-banku-A-more-deadly-substitute-for-cigarettes-1229509.

[17] Omari R, Frempong G. 'Food safety concerns of fast food consumers in urban Ghana,' *Appetite*, 98 (2016): 5.

[18] Jessica Ham, interview.

[19] 'FAQ,' Ghana Food Movement, accessed May 2021, https://www.ghanafoodmovement.com/faq.html.

> Does Ghana have a proper restaurant inspection system? Just after Coming To America movie 2, they all wanna start launching in Africa with all these unhealthy foods. Ghana's Asanka Locals and the rest must be supported heavily. No tourist will visit Ghana because of Dominos Pizza restaurant, but rather a Kontomere with Yam or Banku and Tilapia fast food restaurant. Branding and execution is what Ghana or perhaps Africa needs. It's the 21st century, the technology and the technical know how is right here. Arise Ghana youth for your country![20]

Indeed, this concern with the preservation of local foods and identities extends all the way to the Upper West and Brong Ahafo Regions. The Center for Indigenous Knowledge and Organizational Development (CIKOD) promotes agroecological knowledge by expanding learning about local seed varieties as well as promoting seed saving and sharing. The goal of the organization is to aim for a better understanding of African spirituality in food and farming and the application of African spirituality led by 'traditional wisdom holders.'[21] The activities of CIKOD reveals a deep concern over the threat packaged foods represent to traditional food knowledge and related beliefs.

Not only do some believe that packaged and imported foods imperil regional identity and traditional knowledge, but they also jeopardize Ghana's existence 'as a politically independent people.'[22] As previously mentioned in Chap. 7, Food Sovereignty Ghana (FSG) is an organization whose mission is to protect national control over resources like land and water access. They posit that multinational business models marginalize smallholders by promoting unsustainable farming methods and inequitable food trade. Seeking to go from 'food security to food sovereignty', FSG addresses supply issues, price fluctuations, and policy conditions that come with dependence on food imports. By promoting Ghanaians right to healthy and 'culturally appropriate' food, they hope to ensure that smallholders are sustained by state provisions over foreign

[20] 'Domino's Pizza.'

[21] 'Center For Indigenous Knowledge And Organizational Development (CIKOD),' Seeds, Soil & Culture, accessed May 2021, https://seedssoilculture.org/grants-arch/center-indigenous-knowledge-organizational-development-cikod/.

[22] 'About Us,' Food Sovereignty Ghana, accessed May 2021, https://foodsovereigntyghana.org/about-us/.

multinationals. Among the programs for smallholders is a program to ensure that they have access to necessary infrastructure such as credit, technology, and transport. In sum, then, the debate is over what is of greatest concern to Ghanaians about food consumption: the nutritional health of the populace, the preservation and retention of local, traditional foodways, or the preservation of national economic security. All these issues are in fact related issues with no easy solutions.

Though I concede that Ghana may be experiencing health issues connected to the consumption of highly processed foods I still maintain that the focus upon nutrition—another name for scarcity—fails to address the real needs of Ghanaians in ensuring the integrity of their foodways and their national sovereignty. These public discussions about nutrition and foreign foodways obscure both longstanding issues with transportation infrastructure, as well as the climate issues farmers are experiencing. As stated in Chap. 7, transportation and storage issues have been an issue since the colonial period and are often overlooked today in the general discourse over food.[23] These pressing infrastructure issues persist today if Twitter appeals to the Ghanaian government are to be credited. Since the beginning of May 2021, many have been using social media to send their government a message: #FixTheCountry.[24] They complain about poor roads, *dumsor* (power cuts), and water shortages, all of which inhibit the internal transport and storage of food.

Climate change too is another global issue that is affecting the way food is produced and compromising the economic health of the nation. As discussed in Chap. 6, market forces (urban demand) are increasingly driving cash crop production in the north.[25] Climate change is causing changes to gendered labour practices in the north, forcing women and

---

[23] 'Minutes of the Meeting Held in the Director of Supply's Office', 25 October 1941, ARG 1/1/67, PRAAD, Accra.; Kwame Nkrumah, 'Circular from Prime Minister Kwame Nkrumah to Chief Regional Officers on Behalf of The National Food Board', 1953, ARG 2/8/27, PRAAD, Accra.; 'Transport issues,' The Regional Officer, ARG 2/8/118, 27 April 1977.

[24] 'Does Ghana Really Need Fixing?' Africa Daily, BBC Sounds, accessed May 2021, https://www.bbc.co.uk/sounds/play/p09h6jcq.

[25] Awo, Jessica R. Ham, '"Who Knows Tomorrow?": Food Insecurity, Distress, and Managing the Future in Upper West Ghana.' (PhD diss., University of Georgia, Athens, 2016), 18. Padmanabhan, 'The Making and Unmaking', 66.

young people to migrate to insecure jobs in the south. Similar to what the south experienced in the colonial era, women in the north are struggling to maintain their ability to produce traditional foods because men are taking over the production of some foods that have become cash crops. Also, climate change is affecting the availability of traditional foods like dawadawa, causing a switch over to soybeans. As more men are taking over land traditionally farmed by women for food to farm cash crops, many women and young people are forced to travel to the south to make a living.

Many of these cash crops are increasingly being farmed with off-farm inputs such as genetically modified seeds and chemical fertilizers. Despite the failings of the first Green Revolution in the 1960s, a second wave emerged in the early twenty-first century, this time primarily targeting the African continent. In Ghana, for example, the government worked with donor organisations and the private sector to extend the Green Revolution throughout its major food producing areas. Although new technologies may have increased yields, they have also raised costs of production and the region has no structured market systems that can ensure farmers generate an income from crops. Commercial farmers who can produce in large quantities are often linked to markets through contract buyers who purchase direct from their farms. Poor and small-scale farmers are unable to reap the same rewards due to the expense of costly off-farm inputs and lack of access to ready markets. This overall situation exacerbates the trend whereby farmers are moving away from food crops and towards cash crops for export. These market conditions have already threatened local food systems and will continue to do so.[26] Ghana needs improved transport and targeted assistance with climate change. These are not sexy issues for the West as these issues are resistant to technological fixes. Instead, the focus has been on nutrition and lack and a narrative that Ghana is under attack by industrial food giants of packaged and fast food.

---

[26] James Boafo and Kristen Lyons, 'Ghana's farmers aren't all seeing the fruits of a Green Revolution,' The Conversation, accessed June 2021, https://theconversation.com/ghanas-farmers-arent-all-seeing-the-fruits-of-a-green-revolution-162345.

Current debates over nutrition also have the effect of overlooking the creative ways Ghanaians seek to care for others by cooking and serving regional foods. One example of this creative way of cooking can be seen at Tatale Restaurant in Osu. In a video produced in cooperation with Lotte Wouters of the GFM entitled 'Spicy Ghana Food Tour,' Tatale's proprietor, Chef Ben Asamani, features vegan local dishes, artfully presented.[27] Other menu items include fufu, abolo, and egushi (greens and ground melon seed stew) are artfully presented in an *Haute cuisine* style. GFM continued its promotion of artfully presented regional cuisine with a visit to Northern Platter Restaurant in Accra. Northern Platter serves northern dishes like tubaani (steamed bean and millet dumplings), tuo zaafi, and wagashi, a fresh fried cheese dish representative of the Fulani of Ghana and Benin.[28] These new approaches to cooking have the effect of promoting regional cuisine to both locals and foreign tourists while at the same time preserving local foodways.

Not only are Ghana chefs experimenting with the cooking and presentation of regional foods in western style restaurants, but some are working within the current packaged and fresh food distribution system to curb waste. Chef Elija Amoo Addo, Executive Director of Food 4 All Africa (FFAA) is another example of creativity combined with service. Chef Addo runs the first and largest food bank in Africa.[29] Addo realized that food waste was not just limited to the hospitality industry but was all along the supply chain. The problem of food waste was everywhere from manufacturing companies, upmarket hotels, suppliers, and importers. An importer of corned beef for example may choose to destroy mislabelled food rather than attempt to sell it. FFAA collects surplus food, prepares it and distributes it to vulnerable people in Accra. The aged and disabled receive food boxes. The hot meals program delivers cooked food such as curried rice and chicken twice weekly to the homeless in and

[27] Davidsbeenhere, 'Spicy GHANA FOOD TOUR!! Tatale, Egusi & Abolo,' YouTube, accessed May 2021, https://youtu.be/3D_tgwzwKO0.

[28] Davidsbeenhere, 'Northern GHANA FOOD in Accra!! Tuo Zaafi, Wagashi & Kuli Kuli,' YouTube, accessed May 2021, https://youtu.be/RvY2XwjoX3c.

[29] 'Kicking out food waste, feeding the vulnerable in Ghana—The Food4All way,' General News of Tuesday, 20 April 2021, Ghana Web, accessed May 2021, https://www.ghanaweb.com/GhanaHomePage/NewsArchive/Kicking-out-food-waste-feeding-the-vulnerable-in-Ghana-The-Food4All-way-1237636.

around Accra. FFAA also runs a school feeding program for schools that do not have one. This is a creative way of redistributing both packaged and fresh food in and around Accra.

To be sure, Ghana is experiencing many of the typical characteristics of twentieth to twenty-first century globalisation as defined by David Inglis in his 'Globalisation and Food: The dialectics of globality and locality': increased urbanization, decreased peasantry, larger farm production units, the application of technology to production and consumption, international transport systems, and new modes of food preservations are all evident in Ghana and discussed in the different chapters of this monograph.[30] The only characteristics which are not yet readily evident in Ghana are the mass market livestock production and its attendant ethical issues with the mass killing of animals. As discussed in Chaps. 5 and 7, two of these general responses to globalisation—hybridization and polarization—can be observed in Ghana. Overall, people in Ghana prefer to eat their regional foods, and this is evident across class, age, and ethnic groups.

The overriding process therefore seems to be one of *hybridization* where seasonings and tinned foods are incorporated into the main meal of the day. As discussed in Chap. 3, some foods which could not be incorporated into the main meal were assigned to the breakfast or lunch-time period. What people in Ghana seem to be finding most problematic is the application of technology to food production and the threats to local control over food. Active *polarization* can be seen in anxiety over new technological processes applied to food production that threaten food sovereignty. New organizations which promote regional cuisines and traditional food production knowledge reflect this polarization and anxiety over the convenience, ubiquity, and impersonalisation of fast food, be it glocal (fried rice, jollof rice, fried plantain) or global (KFC, McDonalds). This anxiety manifests itself in questions over food safety of glocal foods, and concerns about diet globalisation which manifests itself in lifestyle diseases such as diabetes or high blood pressure. Since World War I, people in Ghana have experienced issues with the supply of imported and packaged foods largely due to supply issues in other parts of the world.

[30] Inglis, 'Globalization and Food,' 497.

Despite the connectivity of the global food system, its crisis-prone tendencies have prevented a complete homogenization of Ghanaian foodways.[31] Recent discussions of how globalisation affects Ghana tend to focus upon nutritional issues that recall the spectre of lack whenever people talk about Africa and food. This dialogue obscures basic underlying infrastructure and food production issues, the creativity currently being applied to address these issues, and regional specializations that make Ghanaian food interesting and evolving.

This project examined the different ecologies of Ghana and how they structured the availability and choices of starchy staples and their accompaniments. I have shown how these eco-culinary zones signified both collective belonging and otherness through differing regional cuisines. The history of the Atlantic world food exchange cannot be properly understood without a recognition of how African peoples incorporated new foods into their culinary systems whilst maintaining their unique customs. Equally, the retention of cultural values in enduring foodways confirmed that Ghanaians were not helplessly subject to the whims of empire and colonial rule but active participants in the history of the Atlantic world.

The evidence suggests that Ghanaians generally had abundant provisions, which they used to assert the specificity of what they ate. This allowed for defining the otherness of their neighbours, giving meaning to their own place in the world. The changes in foodways over the last two centuries demonstrated how observed traditions could be drastically altered without losing their time-honoured significance. Thus, novelty was steeped in the sauce of tradition and originality was assuaged by the familiar. So long as locals remained at the helm of the variations, they made room within their concepts of the traditional proper meal for innovation and negotiation, protecting the proper meal's cultural continuity. Food systems and their daily expression in commensality practices and in extraordinary food events helped Ghanaians to locate their place in the ecologies in which they found themselves.

At the heart of the way culture and identity was shaped by food is the concept of the proper meal. The historical fashioning of local ethnic

[31] Inglis, 'Globalization and Food,' 498.

identities was realised and reflected in the daily eating practices, in the role of food in myths and legends, and in the manner food was used in ritual and in the building of social and political hierarchies. Chapter 2 examined the synesthetic memory created by the texture of food: its smell, feel, taste and look. These sensory elements combined to construct an embodied experience that was then stored and became part of the cultural memory of different societies and political entities in Ghana. Performative memories were created in the process of re-enacting the skills necessary to make the ideal meal (grinding, pounding, fermentation, etc.) and were reflected in the proper social relationships in the cultural role assigned to women as food preparers.

Chapter 3 analysed further how political power in the great forest kingdom of Asante was embodied by food gifting, feasting and signalling. I argue there was a correlation between the esteem with which different nations were held and the food gifts their envoys received. In a sequence of early nineteenth-century diplomatic encounters, the Asante enacted a process of political negotiations using food and the deprivation of food to broker commerce in the era following the banning of the transatlantic slave trade by the British. From the late 1830s the tenor of feasting changed considerably as the slave trade wound down—presenting a challenge to the economic foundations of the Asante empire. Old relationships between Europeans and the Asante were renegotiated, enabling the Asantehene to preserve a considerable degree of moral authority. Rather than affecting a radical overthrow of Asante institutions, the influence of the Asantehene and his role in Asante cosmology as demonstrated through food in ritual, continued to dominate the religious life of the Asante empire.

Culinary change during the era of transition to legitimate commerce on the Gold Coast was further developed in Chap. 4. Social change, it was argued, was bodily experienced in the staple starches people cooked and ate on the coast. Coastal peoples experienced considerable transformations in established social relationships as a result of the turmoil and economic insecurity of the nineteenth-century transition from the slave trade to the legitimate commerce. A reshaping in the relationships between coastal peoples and European traders affected the terms of commerce, causing a consolidation of intraregional trade in items such as palm oil and fish. Numerous conflicts on the coast initiated by the Asante in the early nineteenth century caused many to turn to cassava, an American import long regarded as the food of slaves and not accepted as part of the local freeborn identity. The new materiality of starches like cassava and rice did not subsume the importance of traditional starches such as millet, which were held in the cultural performative memory via rituals.

Chapter 5 focused on how the concept of the proper meal in the savanna region of northern Ghana differed from the forest and coast to the south. The savanna ecology strongly flavoured culinary culture and the social relationships that flowed from it. The extreme seasonality of the north structured a woman's time in such a way as to centre the need to preserve food for the lean season. Some of the biggest changes in northern Ghana resulted from the movement of people due to the widening influence of global trade on the hinterland and the demand for labourers, first in the palm oil trade from the 1820s, and the later expansion of the cocoa trade and other commercial enterprises on the coast in the late nineteenth and early twentieth centuries. The late twentieth century change to a greater reliance on the market for sustenance and the increased commodification of essential soup ingredients for sale to southern Ghana has altered the relationship of northerners to vegetable and grain production from subsistence to market production. In some ways, these market changes affected the gendered division of labour in food production. Northern Ghanaians have navigated this change by reassigning the gendered connotation of cash crops, or by replicating traditional ingredients by subsuming foreign ingredients into traditional ones.

By the early twentieth century many imports came to be part of the daily food repertoire in the colonial Gold Coast, with tinned fish, tomatoes, and bouillon cubes being among some of the most popular. Chapter 6 traced the reasons for the increase of consumption of imports over time. People expressed agency in the adoption and incorporation of new foods into their quotidian practices of cooking and eating. Some of these imported foods—such as biscuits, sugar, and tea—assisted people in coping with the accelerated, monetised produce export economy by occupying breakfast and lunch mealtimes that had historically been occupied by surplus from the previous evening's meal. Cooks adopted convenience foods to speed the process of preparing the main meal, freeing up time to participate in the growing global export economy.

The reshaping of the relationship with the main meal began to accelerate during the economic distress of the 1970s and 1980s. The need for men and women to work longer hours to make ends meet reoriented women's responsibilities, making the procurement of breakfast and lunch an individual endeavour. Also, the traditional main meal of the day was subject to movement from an everyday event—especially in urban areas—to a weekend event. This renegotiation of women's responsibility to provide the proper meal did make the commercialization of meals and the incorporation of convenience foods earlier in the day more socially acceptable. Despite the reorganization of when and how the proper meal was prepared, gender remained a primary structural factor in the meanings of food. Additionally, Ghanaians readily embraced new foods from immigrants, believing that the variety provided a material and symbolic context for their shared national identity.

Identities in Ghana have been reshaped by a variety of historical factors over the last two hundred years. One of the aspects of identity most significantly impacted has been that of gender. Gender and its associated relations to the means of production situates people and helps them to understand their role and function in society. As Ghana was drawn into the global export economy, gendered relations of production, which were tied to ritual and political systems of power, gradually began to change. The historical transformation from an economy based on farming, fishing and local trade to one increasingly dominated by monetisation and global commercial production caused some women to lose ground in terms of access to the means of production while some men made gains in this same area.

Although some women successfully parlayed service into wealth, many other women toiled in long distribution chains of goods, having lost their access to the means of production. Education, first formally extended to males, became increasingly important as it gave one an advantage in the mediation of access to the means of production and the revenue generated from tariffs. By the early twentieth century, more and more residential dispersions contributed to the stretching of family networks and the weakening of the influence of elders, both male and female. Christianity too became a major social force. The adoption of Christianity in southern Ghana presented challenges to the identities of believers, causing them to construct new identities for themselves with Christian beliefs at the centre. Christianity presented further challenges to family relationships and traditional obligations. Cooking remained a way to honour relationships and obligations and recalled a division of labour that conformed to past expectations. It provided a way for people to look after each other, provided cultural continuity and a connection to place.

At present the literature on African foodways remains relatively generalized to larger regions of Africa and/or is comparative in nature. Furthermore, the food studies scholarship on Africa is oriented toward examining how African foods have impacted the cuisines of people in the Americas. What is required are more specific examinations of how foodways in Africa have changed using local historical knowledge and interpretations. My work contributes to the literature on the cuisine of the Atlantic world by considering how the history and culture of Ghana was closely connected with global trade flows, and could benefit others who are interested in understanding more of the African participation in the Atlantic world. I have learned that local voices are often absent, and the work of African scholars is often not included in the scholarship. This study may contribute to debates over how African history is contested; namely who got to speak in the past and who gets to speak in future.

By engaging with the evidence for Ghana from the nineteenth century to the present, this study has shown an inventive people, evolving within the seasonality of their climates, making tasty food that satisfies the body and the soul. What emerged from this study is evidence of a people dynamically adjusting to change, who had agency with respect to what to eat within the physical structures of their ecological zones, and who chose and continue to choose to honour their sense of place and their respective cultural heritages.

References

I. Adam, S.E. Hiamey, E.A. Afenyo, Students' food safety concerns and choice of eating place in Ghana. Food Contr. **43**, 135–141 (1 September 2014). https://doi.org/10.1016/j.foodcont.2014.03.005

A. Adubia, Changing lives through the internet; the story of a Ghanaian beans seller. The Sun Ghana (blog), 7 April 2021., https://www.thesunghana.com/2021/04/07/changing-lives-through-the-internet-the-story-of-a-ghanaian-beans-seller/

P.K. Akowuah, E. Kobia-Acquah, Childhood obesity and overweight in Ghana: a systematic review and meta-analysis. J. Nutr. Metabol. **2020**, 1907416 (8 April 2020). https://doi.org/10.1155/2020/1907416

R. Aryeetey, A. Lartey, G.S. Marquis, H. Nti, E. Colecraft, P. Brown, Prevalence and predictors of overweight and obesity among school-aged children in urban Ghana. BMC Obes. **4**, 38 (2017). https://doi.org/10.1186/s40608-017-0174-0

Barbara Baëta, Interview by Fran Osseo-Asare. In person at Flair Catering, Cantonments, Accra, 22 March 2018

BBC Sounds, Africa daily—does ghana really need fixing?, https://www.bbc.co.uk/sounds/play/p09h6jcq. Accessed 22 June 2021

J. Boafo, Ghanaians are eating more fast food: the who and the why. The Conversation, http://theconversation.com/ghanaians-are-eating-more-fast-food-the-who-and-the-why-153810. Accessed 22 June 2021

J. Boafo, K. Lyons, Ghana's farmers aren't all seeing the fruits of a green revolution. The Conversation, http://theconversation.com/ghanas-farmers-arent-all-seeing-the-fruits-of-a-green-revolution-162345. Accessed 22 June 2021

Davidsbeenhere, Northern GHANA FOOD in Accra!! Tuo Zaafi, Wagashi & Kuli Kuli | Accra, Ghana. https://www.youtube.com/watch?v=RvY2XwjoX3c. Accessed 22 June 2021

Davidsbeenhere, Spicy GHANA FOOD TOUR!! Tatale, Egusi & Abolo | Accra, Ghana, https://www.youtube.com/watch?v=3D_tgwzwKO0. Accessed 22 June 2021

Duncan-Johnstone, Captain A.C, Minutes of the meeting held in the director of supply's office. Meeting Minutes. Accra: Chief Commissioner Ashanti—Gold Coast Colony. ARG 1/1/67. PRAAD Accra, 25 October 1941

Food and Agriculture Organization of the United States, FAO GIEWS country brief on Ghana, http://www.fao.org/giews/countrybrief/country.jsp?code=GHA. Accessed 22 June 2021

Food Sovereignty Ghana, About us | food sovereignty Ghana, https://foodsovereigntyghana.org/about-us/. Accessed 22 June 2021

Ghana Food Movement, FAQ, https://www.ghanafoodmovement.com/faq.html. Accessed 22 June 2021

GhanaWeb, Domino's pizza, world's largest pizza company, set to launch in Accra, http://www.ghanaweb.com/GhanaHomePage/NewsArchive/Domino-s-Pizza-world-s-largest-pizza-company-set-to-launch-in-Accra-1211608?gallery=1. Accessed 22 June 2021a

GhanaWeb, Kicking out food waste, feeding the vulnerable in Ghana—The Food4All Way, 20 April 2021b., https://www.ghanaweb.com/GhanaHomePage/NewsArchive/Kicking-out-food-waste-feeding-the-vulnerable-in-Ghana-The-Food4All-way-1237636

GhanaWeb, Polythene wrapped Banku: a more deadly substitute for cigarettes?, 11 April 2021c., https://www.ghanaweb.com/GhanaHomePage/NewsArchive/Polythene-wrapped-banku-A-more-deadly-substitute-for-cigarettes-1229509

J.R. Ham, Interview by author. Via Zoom, 21 April 2021

D. Inglis, Globalization and food: the dialectics of globality and locality, in *The Routledge International Handbook of Globalization Studies*, (Routledge, 2011)

R. Jenkins, Globalization, production, employment and poverty: debates and evidence. J. Int. Dev. **16**(1), 1–12 (2004). https://doi.org/10.1002/jid.1059

Modern Ghana, Your eating pattern determines your health and lifespan: including should women fast?, https://www.modernghana.com/news/1073413/your-eating-pattern-determines-your-health-and.html. Accessed 22 June 2021

MyJoyOnline.Com, Akufo-Addo concerned about Ghana's obesity rate, https://www.myjoyonline.com/akufo-addo-concerned-about-ghanas-obesity-rate/. Accessed 22 June 2021

K. Nkrumah, Circular from Prime Minister Kwame Nkrumah to chief regional officers on behalf of the national food board. ARG 2/8/27. PRAAD Accra, 1953

R. Omari, G. Frempong, Food safety concerns of fast food consumers in urban Ghana. Appetite **98**, 49–54 (1 March 2016). https://doi.org/10.1016/j.appet.2015.12.007

G.O. Otchere, SONA 2020: Akufo-Addo expresses worry about obesity in Ghana [Video]. Adomonline.Com (blog), 20 February 2020. https://www.adomonline.com/sona-2020-akufo-addo-expresses-worry-about-obesity-in-ghana-video/

Seeds, Soil & Culture, Center for Indigenous Knowledge and Organizational Development (CIKOD), https://seedssoilculture.org/grants-arch/center-indigenous-knowledge-organizational-development-cikod/. Accessed 22 June 2021

S. Stevano, The urban food question in the context of inequality and dietary change: a study of schoolchildren in Accra. J. Dev. Stud. **56**(6), 1177–1189 (2 June 2020). https://doi.org/10.1080/00220388.2019.1632434

S. Stevano, D. Johnston, E. Codjoe, Better decisions for food security? critical reflections on the economics of food choice and decision-making in development economics. Camb. J. Econ. **44**(4), 813–833 (7 July 2020). https://doi.org/10.1093/cje/beaa012

The Regional Officer, Transport issues. ARG 2/8/118. PRAAD Accra, 27 April 1977

S. Zaman, N. Selim, T. Joarder, McDonaldization without a McDonald's. Food, Cult. Soc. **16**(4), 551–568 (1 December 2013). https://doi.org/10.2752/175174413X13758634982010

Index[1]

[1] Note: Page numbers followed by 'n' refer to notes.

B. Simpson Miller, *Food and Identity in Nineteenth and Twentieth Century Ghana*, Food and Identity in a Globalising World,
https://doi.org/10.1007/978-3-030-88403-1

G

www.ingramcontent.com/pod-product-compliance
Lightning Source LLC
Chambersburg PA
CBHW061337170725
29767CB00003B/18
* 9 7 8 3 0 3 0 8 8 4 0 2 4 *